Fundamentals of Materials Science and Engineering

Fundamentals of Materials Science and Engineering

Editor

Sumit Pandey

scitus
academics

Fundamentals of Materials Science and Engineering

Edited by **Sumit Pandey**

Printed in 2017

ISBN: 978-1-68117-218-7

Library of Congress Control Number: 2015936579

SCITUS Academics LLC,
616, Corporate Way, Suite 2, 4766,
Valley Cottage, NY 10989

www.scitusacademics.com

Notice

Contents

Preface

Fundamentals of Materials Science and Engineering is an alternate version of my text, Materials Science and Engineering: An Introduction, Fifth Edition. The contents of both are the same, but the order of presentation differs and Fundamentals utilizes newer technologies to enhance teaching and learning. With regard to the order of presentation, there are two common approaches to teaching materials science and engineering—one that I call the "traditional" approach, the other which most refer to as the "integrated" approach. With the traditional approach, structures/characteristics/properties of metals are presented first, followed by an analogous discussion of ceramic materials and polymers. Introduction, Fifth Edition is organized in this manner, which is preferred by many materials science and engineering instructors. With the integrated approach, one particular structure, characteristic, or property for all three material types is presented before moving on to the discussion of another structure/characteristic/property.

Editor

Chapter 1

Influence of Partitioning Process on the Microstructure and Mechanical Properties of High Deformability Oil-Gas Pipeline

Jing Ma[1], Xiaoyong Zhang[1], Shixia Cheng[2], and Huilin Gao[1]

[1]Department of Material Science and Engineering, Xi'an Shiyou University, Xi'an, Shaanxi 710065, China

[2]State Key Laboratory for Mechanical Behavior of Materials, Xi'an Jiaotong University, Xi'an 710049, China

ABSTRACT

Multiphase structure of bainite and M/A constituent can be obtained for X80 oil-gas pipeline through a novel heat online partitioning (HOP) technology. The effects of partitioning temperature on the microstructure and mechanical properties of the experimental steels were researched by means of mechanical properties test, microscopic analysis, and X-ray diffraction. The results show that with the increase of the partitioning temperature, the strength of the experimental steel decreases and the ductility increases because of the increase of bainite lath width, the decrease of dislocation density, the increase of retained austenite content, and carbides coarsening. The decrease of the volume content and stability of retained austenite is the key factor, which leads to the increase of strength and the decrease of plasticity in a high range of partitioning temperature.

INTRODUCTION

Extensive studies to develop high deformability line pipe have been conducted. Recent studies [1, 2] indicate the significant economic advantages of using higher grade line pipes in constructing long distance pipelines. On the other hand, construction of pipeline has expanded into the environmentally severe regions such as cold region, seismic region, deepwater, and sour gas environment. Various material properties in addition to higher strength, such as high toughness, high deformability and sour resistance property, are required for recent line pipe steels. One of the most challenging fields for pipeline developments is thought to be the seismic and permafrost regions where large plastic deformation is expected to be introduced to buried line pipes.

In order to improve the deformability of pipes, two different types of microstructural control technologies were proposed, based on theoretical and analytic studies on the effect of microstructural characteristics on stress-strain behavior. Grade X65 to X100 line pipes with ferrite-bainite microstructure were manufactured by optimizing the microstructural characteristics. Grade X80 line pipes with bainitic microstructure containing dispersed fine M/A constituents were also developed by applying new conceptual TMCP process. Steel plates with bainite and martensite-austenite constituent (M/A) microstructure can be obtained by applying heat online partitioning (HOP) process subsequently after accelerated cooling process. HOP process is an induction heating process that heats steel plates rapidly. Variety of microstructural controls, such as fine carbide precipitation and M/A formation, can be utilized by this newly developed heat process. One of the significant features of the HOP process is to improve resistance to strain-aged hardening. Both free carbon and dislocation density can be reduced by carbides precipitation and tempering of bainite during the partitioning step of HOP process [3–6]. Therefore, the partitioning process plays a key role in improving the deformability of pipeline steels. Trial production of X80 pipeline steel was conducted by applying the HOP process. Microstructural characteristics and mechanical properties of developed X80 high deformability line pipes are introduced in this paper.

MATERIAL AND METHOD

Trial production of grade X80 pipeline steel was conducted by applying the heat online partitioning process. Table 1 shows the chemical composition of grade X80 pipeline steel. Plates with the thickness of 18.4 mm were manufactured in the plate production mill, which is equipped with accelerated cooling and online heating devices. Plate manufacturing conditions in controlled rolling, accelerated cooling, and heat treatment online process were precisely controlled so that enough strength and high deformability were achieved.

Table 1: Chemical composition of the trial X80 pipeline steel

Grade	Chemical composition (mass %)				Others
	C	Si	Mn	Mo	
X80	0.049	0.23	1.80	0.28	Al, Cr, Ni, Cu, V, Nb, Ti

The heat online partitioning process was developed in order to produce various properties in addition to higher strength, such as toughness, deformability, and sour resistant property [7, 8]. Figure 1(a) shows the schematic illustration of temperature profiles in plate manufacturing process; Figure 1(b) is Gleeble 3500 thermal simulator. The HOP process divides into three steps. First, accelerated cooling (ACC) is stopped above the bainite transformation finishing temperature, where untransformed austenite remains. At this stage, the microstructure is bainite and untransformed austenite. Secondly, immediately after ACC, partitioning at 410°C, 440°C, 470°C, and 500°C (TP = 410~500°C for 500 s), respectively. The subsequent step is partitioning above the fast quenching temperature Tq to accomplish the partitioning (diffusion) of carbon from supersaturated bainite into retained austenite so as to keep carbon enriched austenite stable during subsequently cooling to room temperature. During the heating, carbon in bainite diffuses into austenite. Finally, after the quenching, austenite with higher carbon content retained, and this can transform into MA by water cooling because of highly concentrated carbon in the austenite.

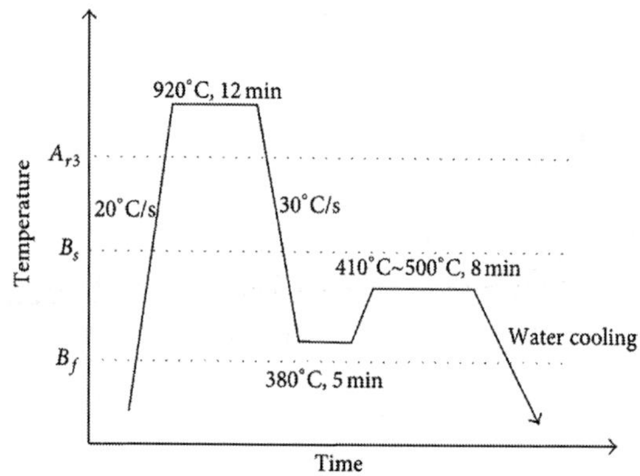

(a)

(b)

Figure 1: HOP thermal simulation process. (a) Curve and parameters (b) Gleeble 3500 thermal simulator.

Volume fraction of MA is affected by chemical composition of the steel and ACC and heating conditions. Effect of volume fraction of MA on Y/T ratio of steel plates which was applied with the new process was studied [9, 10]. Y/T ratio decreased with the partitioning temperature increasing, and Y/T ratio below 85% was able to be achieved by MA volume fraction above 5%. The studies indicated that the heat online partitioning is a promising way to produce high performance pipeline materials that can fulfill various requirements for recent pipelines.

Tensile properties were analyzed in accordance with the ASTM A370 method. The tensile specimens were machined with their axis oriented parallel to the rolling direction and cut into the diameter of 5 mm, the criterion length of 25 mm, shown in Figure 2. They were tested on a MTS810 machine at the strain rate of 10^{-3} m/m·s^{-1} at room temperature. Charpy impact test was performed on standard Charpy V-notch specimens (size: 10 × 10 × 55 mm) along the thickness direction of the steel body at 253 k temperature (−20°C) in accordance with the ASTM E23 by a Charpy impact tester of 500 J capacity, as shown in Figure 3.

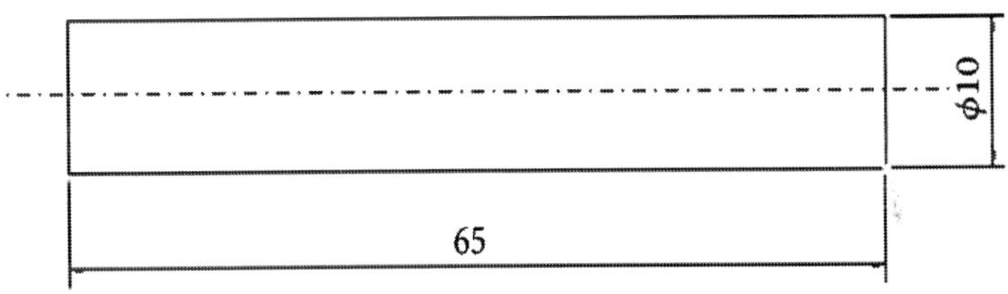

(a)

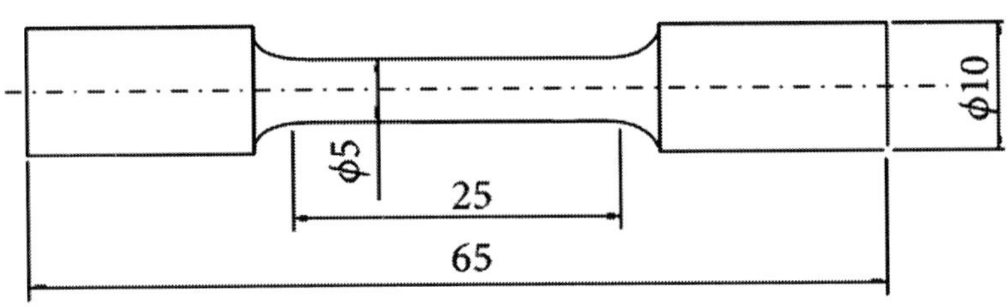

(b)

Figure 2: Tensile specimens. (a) Original size (b) Gauge size.

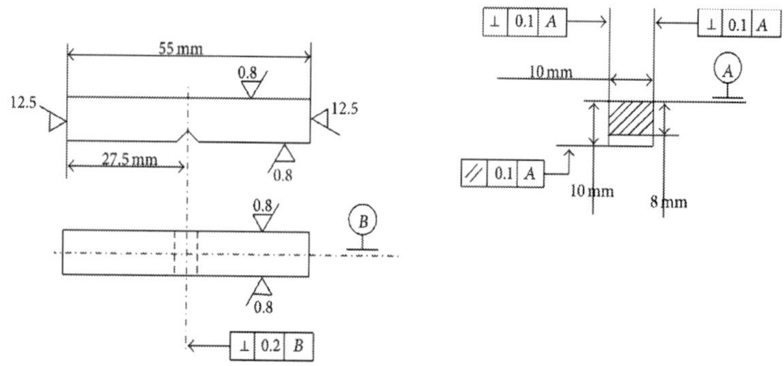

Figure 3: Charpy V-notch specimens.

As Figure 4 suggests, the microstructural examination of the sample was conducted using JSM-6390A scanning electron microscope (SEM) after conventional nital etching. Transmission electron microscope (TEM) specimens were first mechanically thinned to 0.5 mm from thin slices, followed by electropolishing in a twin-jet polisher using 4% perchloric acid solution at ~20 mA, 15 V, and 20°C. TEM examination was carried out in a JEM-200CX microscope at 100 kv. Retained austenite contents were determined using an X-ray diffraction technology. XRD measurement was performed using a copper X-ray source operating at 40 Vkv and 150 mA with a nickel filter; scans were run between 40 and 95 deg 2-theta to capture the significant bainite and austenite diffraction peaks.

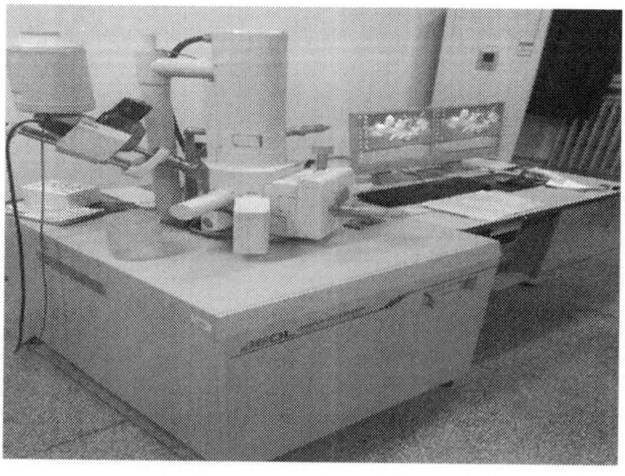

(a)

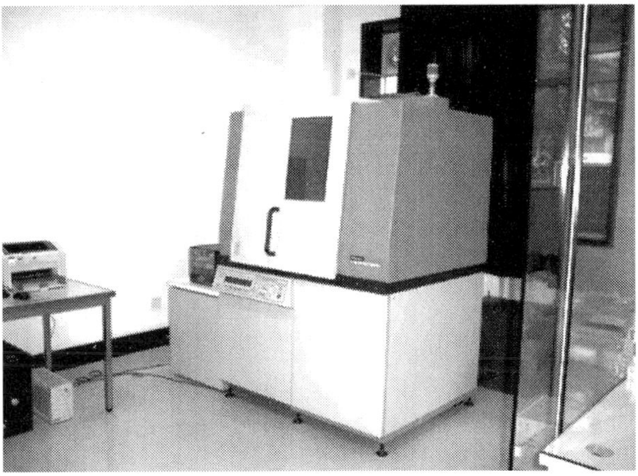

(b)

Figure 4: Microscopic analysis and X-ray diffraction.(a) Scanning electron microscope (SEM) (b) X-ray diffractometer.

RESULTS AND DISCUSSION

Mechanical Properties

X80 grade pipeline steel was developed by applying the heat online partitioning subsequently after accelerated cooling process. Figure 5 shows the engineering stress-strain curves of the trail grade X80 pipeline steels at different partitioning temperature. All pipes show round-house type stress-strain curves. One significant feature of these pipes is the higher levels of uniform elongation.

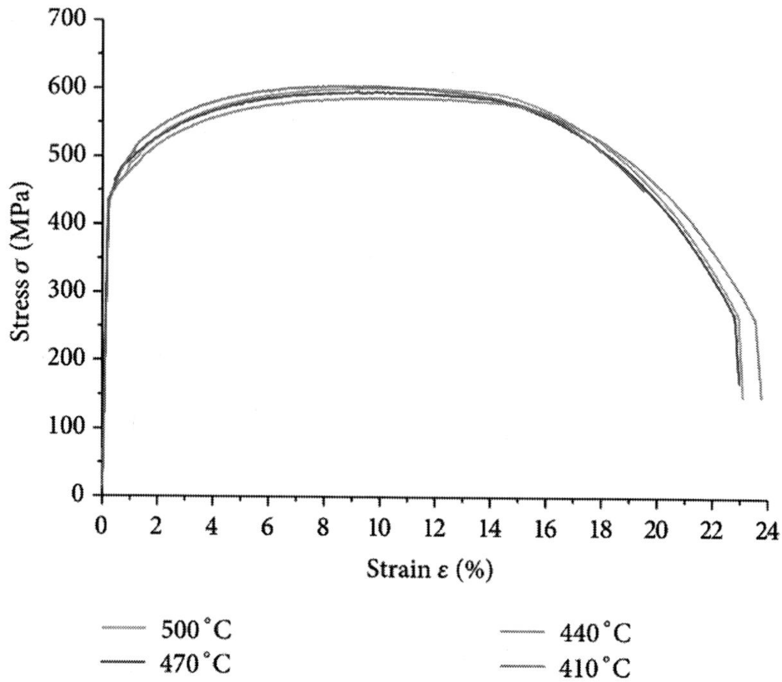

Figure 5: Stress-strain curves of developed X80 line pipes by different partitioning temperature process.

The variation of yield and tensile strength in the range of partitioning temperature (410–500°C) is shown in Figure 6(a). In the experimental steels, the strength indicates the decline trend with the increasing partitioning temperature; while the partitioning temperature is 500°C, the strength increases. However, the effect of partitioning temperature on the plasticity of the dual-phase steels is contrary to the strength trend, shown in Figure 6(b). The trend of variation of the ductility parameters (A, UA) with increasing partitioning temperature is similar for trail steels.

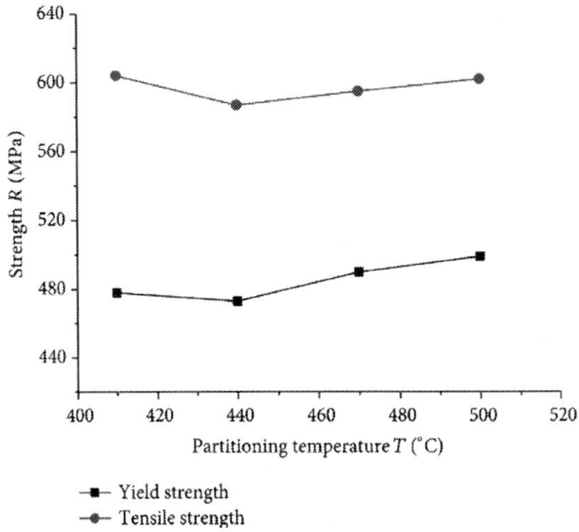

(a)

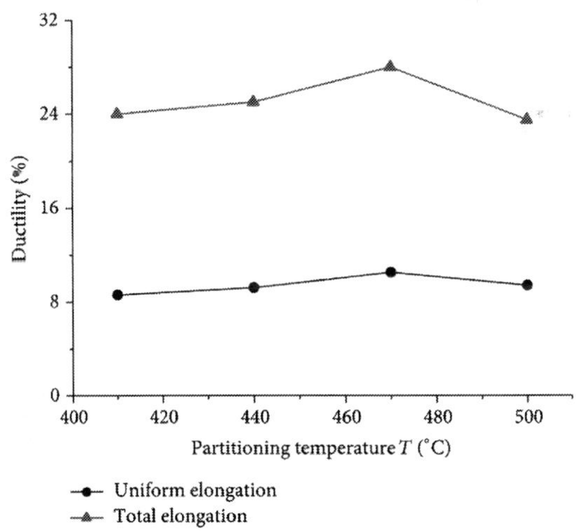

(b)

Figure 6: Relation between partitioning temperature and mechanical properties in tensile test. (a) Strength (b) Ductility.

Table 2 shows the deformability parameters of the trial X80 pipes. The Y/T ratio decreases with the increasing partitioning temperature. All experimental steels at different partitioning temperature show low Y/T ratio, high uniform elongation, and strain hardening exponents. From these results, it is considered that developed dual-phase line pipes with bainite and MA microstructure give significant benefit in high strain applications [10, 11].

Table 2: The yield ratio, uniform elongation, and strain hardening index of X80 experimental steels at different partitioning temperature

Partitioning temperature/°C	410	440	470	500
$Rt_{0.5}/R_m$	0.83	0.82	0.81	0.79
UA/%	8.6	9.2	10.5	9.4
n	0.13	0.12	0.13	0.12

The Charpy impact toughness values for the trail steels at 273 K (−20°C) temperature are given in Figure 7. As seen in Figure 7, the impact toughness for each sample is more than 300 J. The average value fully satisfies the specifications of ISO for the energy necessary to arrest propagation of unstable ductile fracture of X80 line pipes.

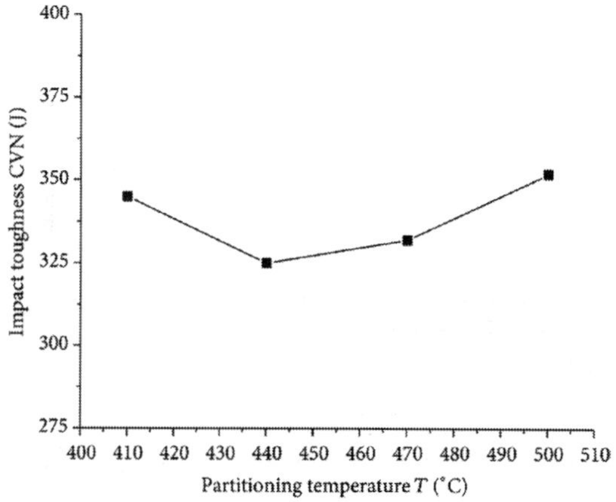

Figure 7: Impact toughness of X80 experimental steel at different partitioning temperature.

Microstructure

Figure 8 presents the evolution of microstructure in the trail steels treated by different partitioning temperature. It is clear that the microstructural feature of these samples is typical bainite (dark field) with fine dispersed MA (bright field).

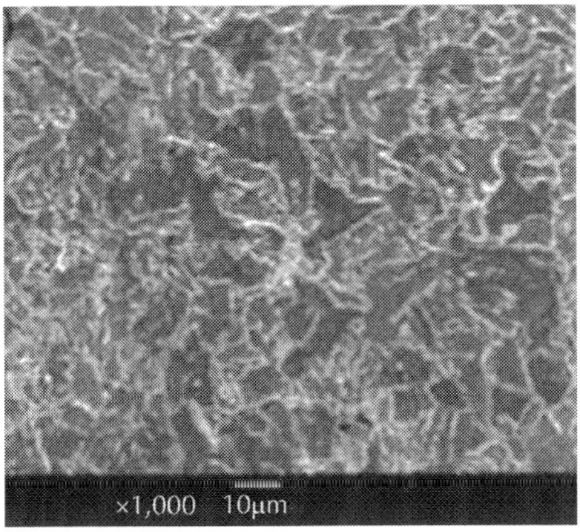

(a)

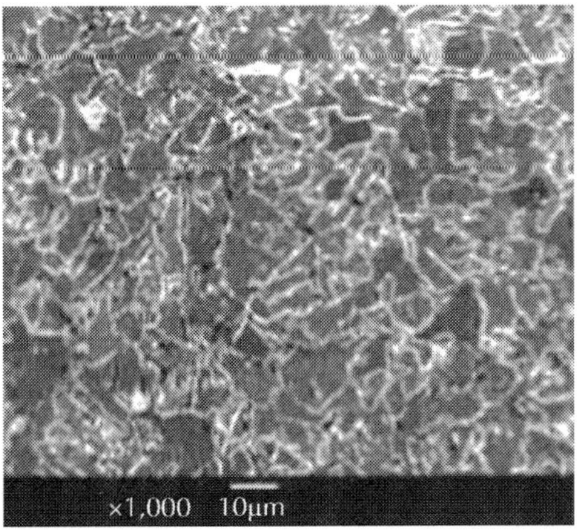

(b)

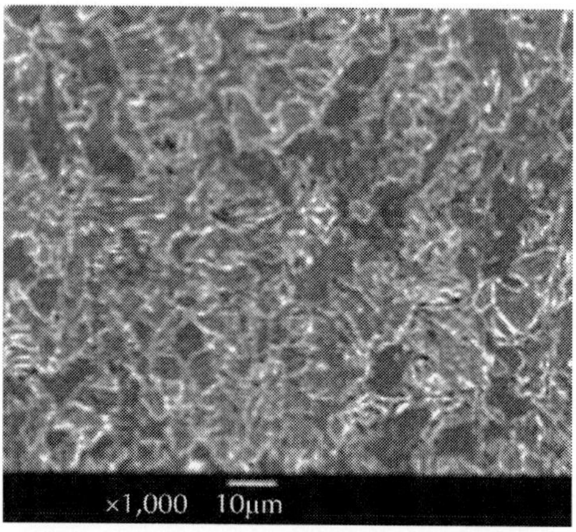

(c)

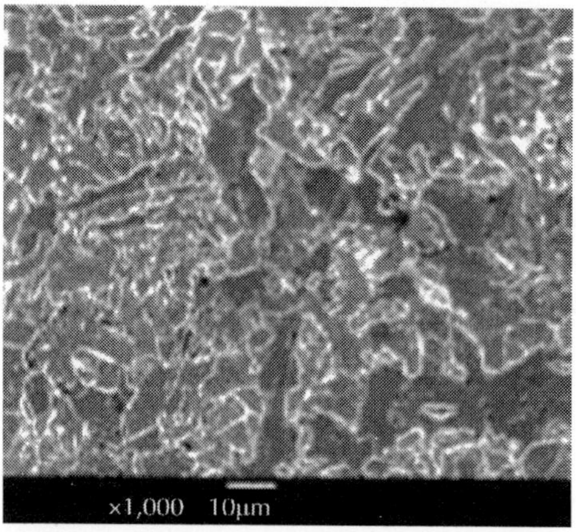

(d)

Figure 8: SEM micrograph of X80 samples at different partitioning temperature.(a) 410°C (b) 440°C (c) 470°C (d) 500°C.

As shown in Figure 8, the experimental X80 steel has similar microstructure characteristic. However, the difference is that untransformed austenite maybe transformed into M/A with adequate partitioning time when the partitioning temperature is above 440°C.

At the same time, higher partitioning temperature promotes carbides precipitation from supersaturated bainite and then maybe transform bainite into the polygonal ferrite. Examples of these microstructure features are indicated with arrows in Figures 9(a) and 9(b). Nanometer-sized fine carbide precipitates were observed.

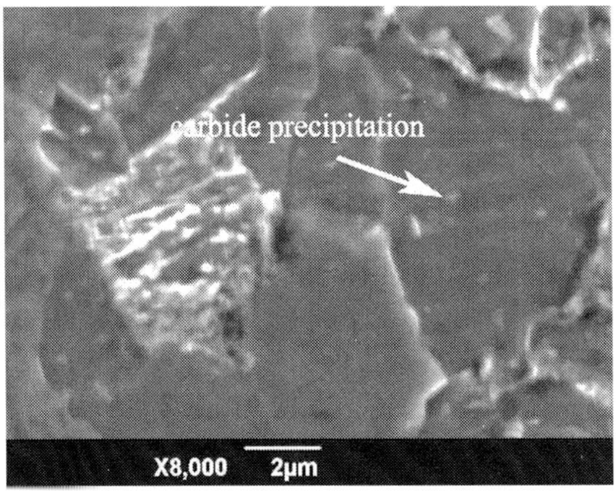

(a)

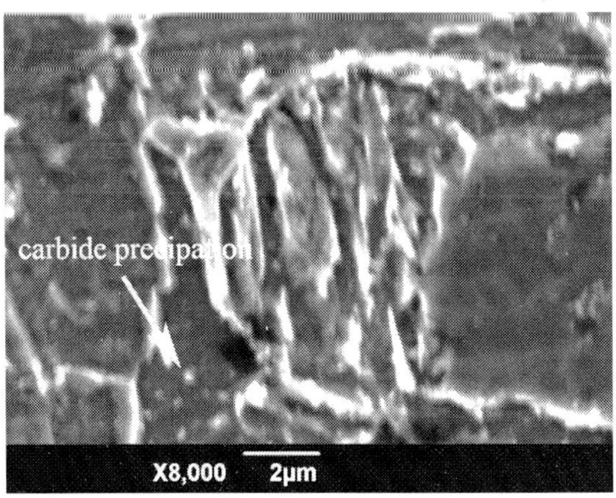

(b)

Figure 9: TEM micrograph of carbides precipitation at higher partitioning temperature. (a) 470°C (b) 500°C.

Discussion

Based on the above results on microstructure and mechanical properties, high strength pipeline steels with lower yield ratio, higher uniform elongation, and strain hardening index were developed aiming to meet the technical requirements of high deformability line pipes and are suitable for the environmentally severe regions such as permafrost and seismic regions. It is well known that the good contribution of high strength and ductility of trail steels is mainly due to the various changes of (B + M/A) dual-phase microstructure.

Bainite. Representative TEM micrograph illustrating the microstructure of X80 pipeline steels treated by partitioning at 653 K and 773 K is presented in Figure 10. It can be also seen from Figure10 that the typical bainite laths and dislocation structures are observed in the bainite matrix of all samples.

(a)

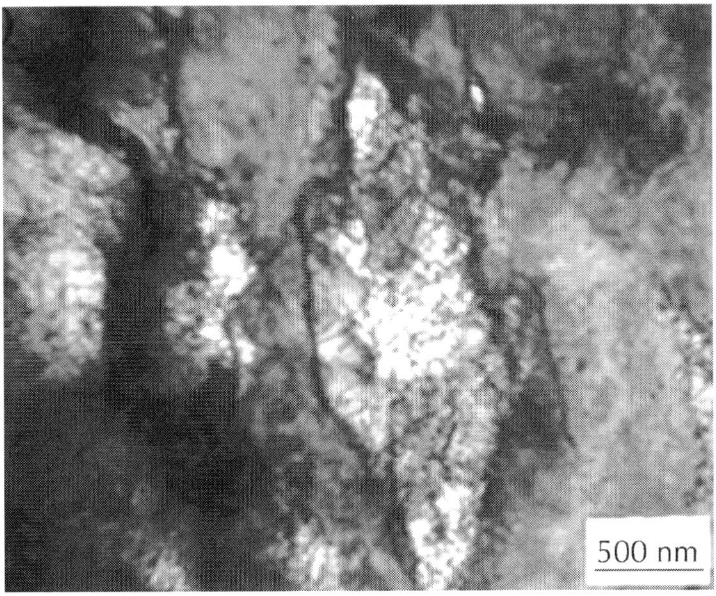

(b)

Figure 10: Typical TEM micrograph of X80 samples treated at different partitioning temperature. (a) 410°C (b) 500°C.

Figure 10 shows the morphology change of bainite laths with the increase of the partitioning temperature. In the high-temperature partitioning process, the lath structure of samples is not obvious because the migration or disappearance of bainite/austenite interfaces and subboundaries makes the bainite laths confused. On the other hand, adjacent bainite laths with small angle orientation may migrate and fuse, which results in the coarsening of bainite laths [12, 13]. With the further increase of partitioning temperature, it can be observed that a part of the bainite is transformed into polygonal ferrite for the recovery and recrystallization of bainite. At the same time, the dislocation configuration of bainite changes at the different partitioning temperature, as shown in Figure 11. It is apparent that the samples partitioned at 410°C have higher dislocation density and the samples partitioned at 470°C have lower dislocation density. Therefore, the strength of test steels will decrease while the plasticity is a rise trend with the partitioning temperature increasing.

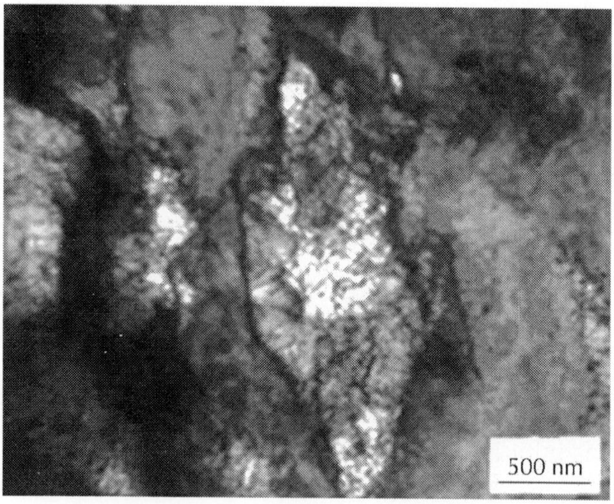

(a)

(b)

Figure 11: TEM micrograph of dislocation configuration in bainite matrix treated at different partitioning temperature.(a) 410°C (b) 470°C

Martensite-Austenite Constituent. One of the most significant features of the heat online partitioning process is to produce dual-phase microstructure which consists of two parts: bainite matrix phase and dispersed MA (martensite-austenite constituent). Figure 12 shows the example of SEM microphotograph of MA produced by HOP process. A large number of MA constituents which are seen as white region were observed in the HOP experimental steels. When the volume fraction of MA was about 13.3%, X80 experimental steels show higher uniform elongation and n-value. Besides, the retained austenite of MA is beneficial to toughness of pipeline steels. Research results show that the high strength of HOP steels results from dislocation-type bainite laths and dispersively distributed carbides in bainite matrix, while excellent plasticity is attributed to the considerable retained austenite [13–15].

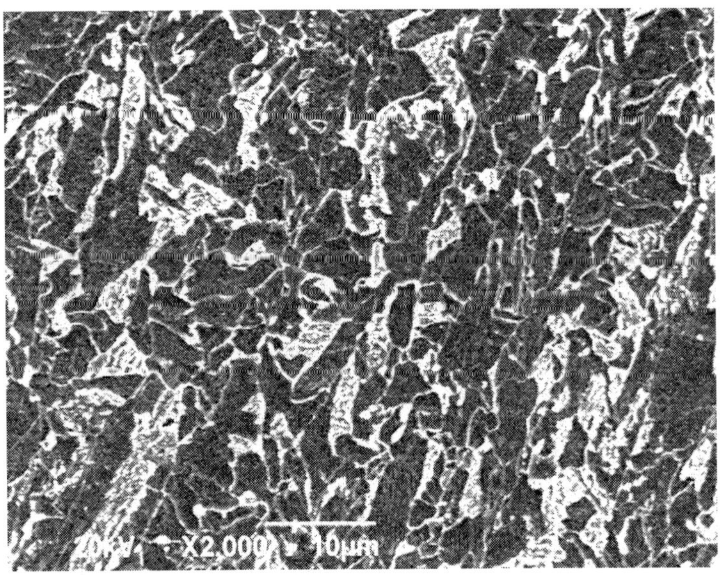

Figure 12: The SEM microstructure of X80 experimental steel with excellent deformability.

Figures 13–15 show the TEM micrograph of (B + MA) dual-phase microstructure. It can be seen that most of retained austenite as thin film morphology were kept treated at low partitioning temperature in Figure 13. The thickness of this retained austenite varies from 50 nm to 100 nm, implying the higher stability of thin film-like austenite than that of the large sized austenite which eventually is changed into fresh martensite after quenching. After short time partitioning treatment, the martensitic structure still remained in the final microstructure and most amounts of carbon atoms are enriched in dislocation and lath boundaries. At the same time, a small proportion of carbon atom diffuses from bainite into untransformed austenite, which results in a small proportion of retained austenite with enough carbon content which can be stabilized at room temperature. With the increase of partitioning temperature, the retained austenite is presented as rods located in the bainite lathes in Figure 14. Therefore, the content of retained austenite increased. At a higher partitioning temperature, the driving force for martensite formation is available. On the other hand, diffusion controlled processes may be activated such as carbides precipitation or further growth of bainite laths by interfaces migration. A different trend is observed for the grade X80 pipeline steel, shown in Figure 15, where the highest partitioning temperature results in the lowest volume fractions of retained austenite. Hence, all diffusion controlled processes resulting in austenite decomposition seem to be promoted.

(a)

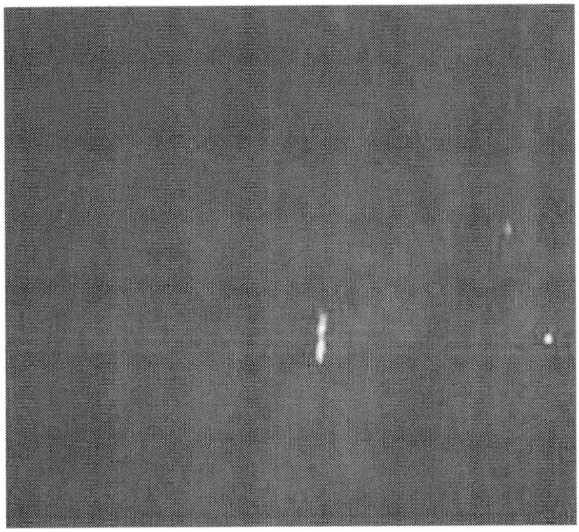

(b)

(c)

Figure 13: Diffraction analysis of retained austenite partitioned at 410°C. (a) Bright field (b) Dark field (c) Electron diffraction spot and indexing.

(a)

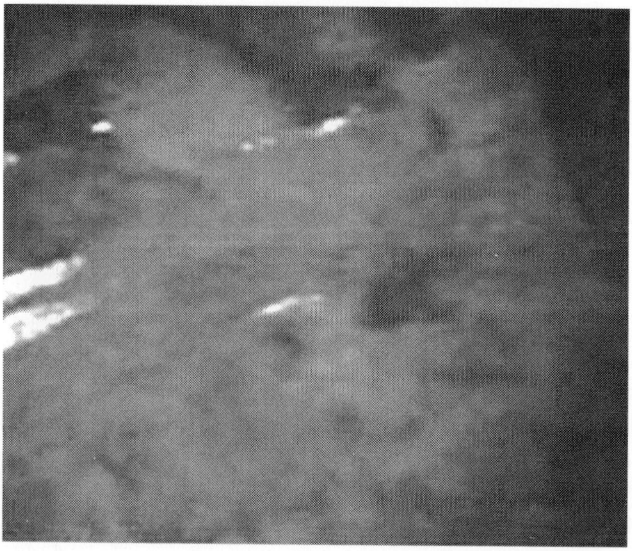

(b)

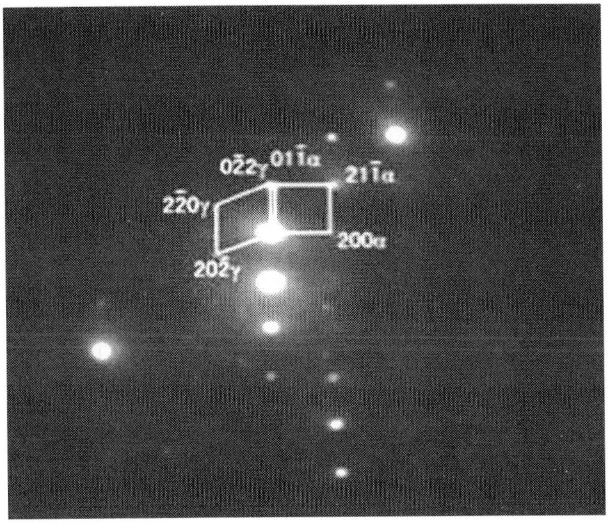

(c)

Figure 14: Diffraction analysis of retained austenite partitioned at 440°C. (a) Bright field (b) Dark field (c) Electron diffraction spot and indexing.

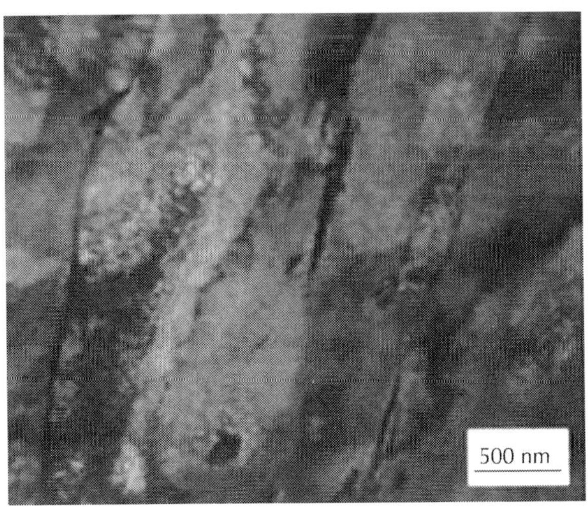

(a)

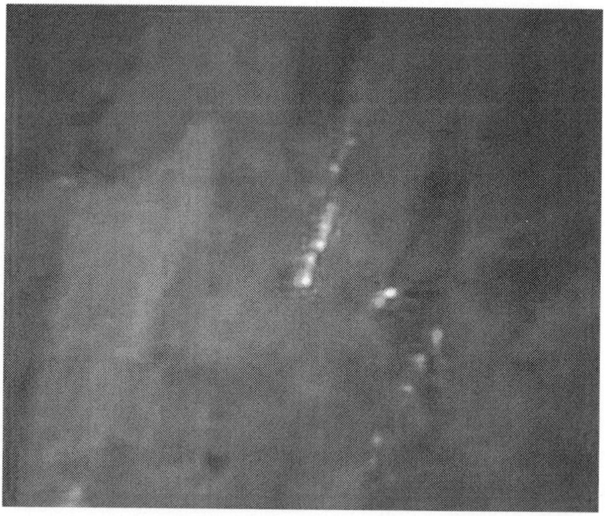

(b)

(c)

Figure 15: Diffraction analysis of retained austenite partitioned at 500°C. (a) Bright field (b) Dark field (c) Electron diffraction spot and indexing.

The X-ray diffraction spectrums of four samples were shown in Figure 16(a), and the retained austenite volume fraction was calculated according to YB/T 5338-2006; the results were shown in Figure 16(b).

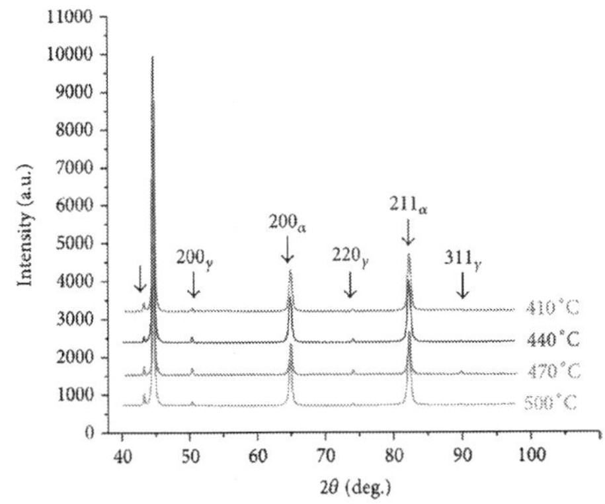

(a)

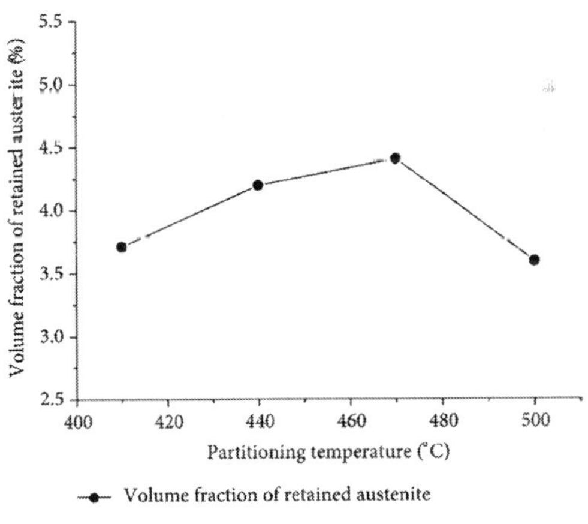

(b)

Figure 16: XRD diffraction spectrums and content of the retained austenite for X80 steels at different partitioning temperature. (a) XRD diffraction spectrum (b) Calculated results about retained austenite.

As shown in Figure 16(b), the variation of volume fraction of presents the peak change with the increasing of partitioning temperature, so did the plasticity. Besides, the stability of retained austenite is determined by carbon concentration. Partitioning at higher temperature is beneficial to the carbons diffusing into austenite from bainite. So from Figure 16(b) we could see that the highest retained austenite volume fraction is obtained when partitioning temperature is 470°C. When the partitioning temperature is enough high, the slope in the change of the retained austenite curve becomes lower owing to the multiple reactions that may occur, such as transition carbides formation, cementite formation, martensite formation, and tempering of bainite, which lead to the decrease in the amount of retained austenite stabilizing at room temperature.

Carbide Precipitation. In order to further identify the microstructures, TEM observation was performed on three typical HOP specimens at 684 K, 731 K, and 773 K (410°C, 440°C, and 500°C). Figure 17 shows the microstructural features of the carbides precipitation of the experimental steel. The formed microalloy carbon and nitrogen compounds during the partitioning process and bainite matrix are harmonious. It is generally believed that transitional ε-carbide dissolved at 400°C and thus mainly formed θ-carbide in the partitioning process. However, sometimes ε-carbide can also be observed in the pipeline steel because of microalloying elements Nb and Mo in experimental steel [16, 17]. As the partitioning temperature proceeds (470 to 500°C), the slope in the change of the retained austenite curve becomes higher and the quantity of θ-carbide precipitation increases gradually, which indicate that θ-carbide formation consumes a large amount of carbon atoms, narrowing the gap in the carbon potential between the bainite and the untransformed austenite. In addition, the migration of carbon atoms from bainite to austenite further leads to very little increase in the amount of retained austenite stabilizing at room temperature.

(a)

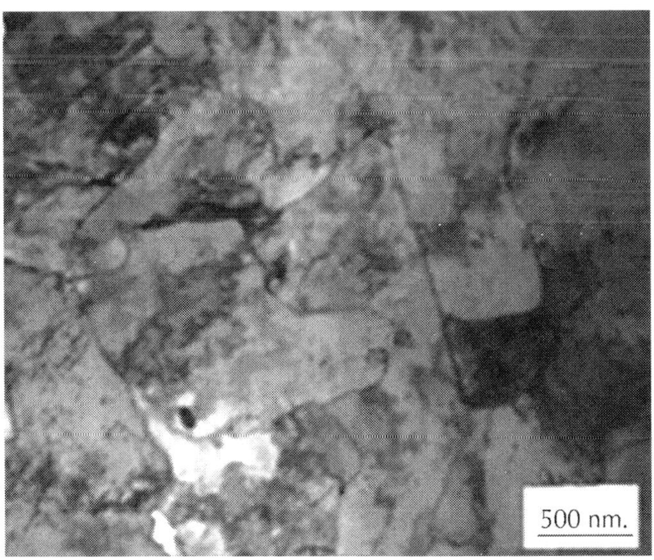

(b)

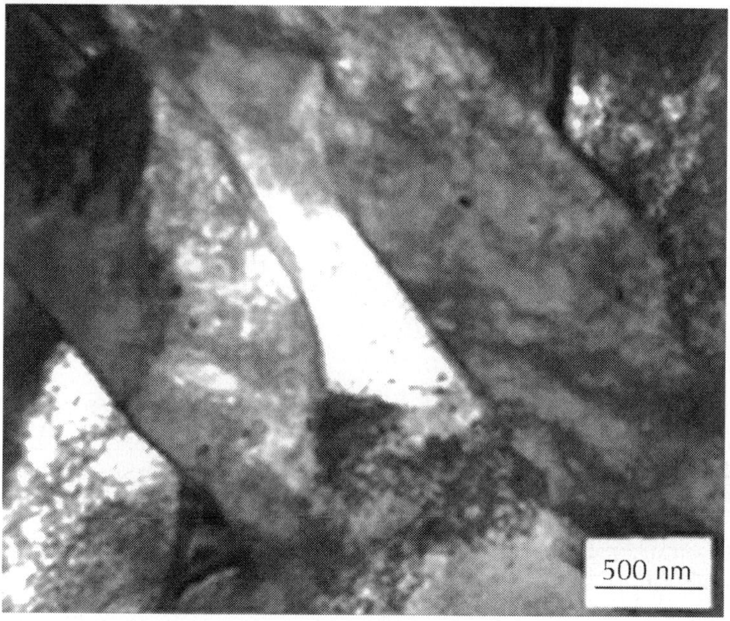

(c)

Figure 17: TEM micrograph of carbides at typical partitioning temperature. (a)410°C (b)440°C (c)500°C.

In general, the weak precipitation hardening effect of cementite and the softening of bainite matrix lead to the poor mechanical properties of HOP specimens during partitioning at 500°C. Besides, TEM observation also indicates that the addition of alloying element Si in X80 pipeline steels can only delay the precipitation of Fe_3C in bainite matrix at relatively low temperature but cannot suppress the formation of Fe_3C at relatively high temperature [18–20]. Analysis results suggest that the formation of these carbides reduces the carbon content in bainite. There is not much of a reduction in the strength of steels, but it can actually increase for steels during the early stages of ε-carbide precipitation. With the partitioning temperature further extended, the carbon content and dislocation density in bainite are reduced, which result in the increase of strength and decrease in plasticity of steels.

CONCLUSIONS

Trial production of grade X80 pipeline steel was conducted by applying the heat on-line partitioning (HOP) process. Microstructure and mechanical properties of the trial X80 pipeline steels and the effect of partitioning process on (B + M/A) X80 were investigated. Results are summarized as follows.

- Trial produced grade X80 pipeline steels showed sufficient strength and excellent deformability for the bainite-MA dual-phase microstructure control.
- With the increasing partitioning temperature, the strength of the experimental steels decreases and the plasticity increases because of the increase of bainite lath width, the decrease of dislocation density, the increase of retained austenite content, and carbides coarsening.
- When the partitioning temperature is enough high, the decrease of the volume content and stability of retained austenite is the key factor influencing the increase of strength and the decrease of plasticity.
- The (B + M/A) X80 steel has lower yield ratio, higher uniform elongation and strain hardening index at different partitioning temperature, aiming to meet the technical requirements of high deformability line pipe.

ACKNOWLEDGMENTS

This work was supported mainly by the National Natural Science Fund of China (Grant no. 51174165). The authors also give great acknowledgments and thanks to the Department of Material Science and Engineering at Xi'an Shiyou University for support of their work.

REFERENCES

1. A. Fujibayashi and K. Omata, Steel's Advanced Manufacturing Technologies for High Performance Steel Plates, JFE Giho, 5th edition, 2004.

2. N. Ishikawa, M. Okatsu, S. Endo, and J. Kondo, "Design concept and production of high deformability linepipe," in Proceedings of the 6th International Pipeline Conference (IPC '06), pp. 215–222, Calgary, Canada, September 2006. · ·
3. T. Shinmiya, N. Ishikawa, M. Okatsu, S. Endo, N. Shikanai, and J. Kondo, "Development of high deformability linepipe with resistance to strain-aged hardening by heat treatment on-line process," in Proceedings of the 16th International Offshore and Polar Engineering Conference, pp. 2963–2968, Lisbon, Portugal, July 2007.
4. H. L. Li, X. Li, and L. K. Ji, "Strain-based design for pipeline and development of pipe steels with high deformation resistance," Welded Pipe and Tub, vol. 30, no. 5, pp. 5–11, 2007.
5. H. L. Gao, Pipeline and Pipeline Steel, China Petrochemical Press, Beijing, China, 2012.
6. A. Glover, "Design, Application and Installation of an X100 Pipeline," in Proceedings of the 22nd International Conference on Offshore Mechanics and Arctic Engineering (OMAE ‹03), p. 37429, Cancun, Mexico, June 2003.
7. N. Ishikawa, S. Endo, and M. Kurihara, "Microstructural aspect of deformability of high strength linepipe steels," Transaction of Material Research Society of Japan, pp. 3469–3472, 2006.
8. S. Zhu, Y. L. Kang, and S. C. Li, "Effect of partitioning temperature on microstructure and mechanical properties of Q&P steel," Transcation of Materials and Heat Treatment, vol. 40, no. 24, pp. 176–178, 2011.
9. D. V. Edmonds, K. He, M. K. Miller et al., "Microstructural features of "quenching and partitioning": a new martensitic steel heat treatment," Materials Science Forum, vol. 539–543, no. 5, pp. 4819–4825, 2007. · ·
10. N. Ishikawa, N. Shikanai, and J. Kondo, "Development of ultra-high strength linepipes with dual-phase microstructure for high strain application," JFE Technical Report, no. 12, pp. 15–19, 2008.
11. J. G. Speer, D. V. Edmonds, F. C. Rizzo, and D. K. Matlock, "Partitioning of carbon from supersaturated plates of ferrite, with application to steel processing and fundamentals of the bainite

transformation," Current Opinion in Solid State and Materials Science, vol. 8, no. 3-4, pp. 219–237, 2004. · ·

12. H.-T. Jiang, D. Tang, Z.-L. Mi, and B.-T. Zhuang, "Effect of partitioning parameters on the retained austenite in low-carbon Q&P steel," Material Science and Technology, vol. 19, no. 1, pp. 100–103, 2011.
13. L.-S. Chen, D. Yang, Y.-Q. Tian, J.-Y. Song, Y.-L. Wei, and Y. Zhao, "Effect of partitioning temperature on microstructure and mechanical properties of low carbon high strength Q&P steel," Transactions of Materials and Heat Treatment, vol. 35, no. 1, pp. 55–59, 2014.
14. P. J. Gibbs, E. de Moor, M. J. Merwin, B. Clausen, J. G. Speer, and D. K. Matlock, "Austenite stability effects on tensile behavior of manganese-enriched- austenite transformation-induced plasticity steel," Metallurgical and Materials Transactions A: Physical Metallurgy and Materials Science, vol. 42, no. 12, pp. 3691–3702, 2011. ·
15. J. Speer, D. K. Matlock, B. C. De Cooman, and J. G. Schroth, "Carbon partitioning into austenite after martensite transformation," Acta Materialia, vol. 51, no. 9, pp. 2611–2622, 2003. · ·
16. Y. Wang, K. Zhang, Z. Guo, N. Chen, and Y. Rong, "A new effect of retained austenite on ductility enhancement of low carbon Q-P-T steel," Acta Metallurgica Sinica, vol. 48, no. 6, pp. 641–648, 2012. · ·
17. Y. Ren, C. Shang, H. Zhang, S. Yuan, and E. Chen, "Effect of retained austenite on toughness and plasticity of 0.23C-1.9Mn-1.6Si steel," Chinese Journal of Materials Research, vol. 28, no. 4, pp. 274–280, 2014.
18. A. J. Clarke, J. G. Speer, D. K. Matlock, F. C. Rizzo, D. V. Edmonds, and M. J. Santofimia, "Influence of carbon partitioning kinetics on final austenite fraction during quenching and partitioning," Scripta Materialia, vol. 61, no. 2, pp. 149–152, 2009. · ·
19. Q. Yu, Y. Sun, H. Ni, and K. Zhang, "Effect of different bainitic microstructures on the mechanical properties of low-carbon steel," Journal of Mechanical Engineering, vol. 45, no. 12, pp. 284–288, 2009. · ·

20. S.-Y. Hao, H.-L. Gao, K.-J. Yan, X.-Y. Zhang, L.-K. Ji, and W.-W. Li, "Microstructure and mechanical properties of x80 pipeline steel with excellent deformability," Journal of Materials Engineering, vol. 65, no. 3, pp. 61–65, 2012.

Chapter 2

Wood versus Plant Fibers: Similarities and Differences in Composite Applications

Bo Madsen[1] and E. Kristofer Gamstedt[2]

[1]Department of Wind Energy, Section of Composites and Materials Mechanics, Risø Campus, Technical University of Denmark, 4000 Roskilde, Denmark

[2]Department of Engineering Sciences, Ångström Laboratory, Uppsala University, 75121 Uppsala, Sweden

ABSTRACT

The work on cellulose fiber composites is typically strictly divided into two separated research fields depending on the fiber origin that is, from wood and from annual plants, representing the two different industries of forest and agriculture, respectively. The present paper evaluates in parallel wood fibers and plant fibers to highlight their similarities and differences regarding their use as reinforcement in composites and to enable mutual transfer of knowledge and technology between the two research fields. The paper gives an introduction to the morphology, chemistry, and ultrastructure of the fibers, the modeling of the mechanical properties of the fibers, the fiber preforms available for manufacturing of composites, the typical mechanical properties of the composites, the modeling of the mechanical properties with focus on composites having a random fiber orientation and a non-negligible porosity content, and finally, the moisture sensitivity of the composites. The performance of wood and plant fiber composites is compared to the synthetic glass and carbon fibers conventionally used for composites, and advantages and disadvantages of the different fibers are discussed.

INTRODUCTION

Composites based on cellulose fibers from wood and plants constitute a relatively new and promising class of composite materials [1–4]. They are environmentally friendly, and they offer good technical performance. For several load-carrying applications, where glass or carbon fiber composites are conventionally used, cellulose fiber composites can be a worthwhile alternative. This is particularly the case for applications where the green advantages (renewability, biodegradability) play an important role, and top-end mechanical properties are not the primary motivation. A vast amount of scientific literature on cellulose fibers for composite applications has been compiled during the last decade (e.g., see recent reviews [5–8]), although the publications tend to be divided into two separate fields depending on the origin of the fibers, that is, from wood or annual plants. The reason for this division is perhaps that the raw materials producers are looking for new markets for their fibers (technology pull), and that the end-users (market pull) have yet to exploit the potential of cellulose fibers, independent of the origin of the fibers. The raw materials producers in this case, that is, forestry for wood fibers and agriculture for plant fibers, have developed their specific technologies along the value chain to produce fibers depending on the traditional usage of the fibers. For wood fibers, pulp mills have been built to produce raw materials for making paper and board. For plant fibers, textile technologies are refined to produce yarns and fabrics. In view of the maturing research field of cellulose fibers shifting towards achieving the technical performance demands of the end-users, and the rather independent existence of research communities of wood and plant fibers, respectively, this paper has been written to shed some further light on the similarities and differences of these two types of cellulose fibers (wood and plant origin), with regard to industrial usage to produce cellulose fiber composites for structural applications. The performance of the fibers will also be compared to the main current competitors, that is, composites made from synthetic fibers, such as glass and carbon fibers. Similarities and differences of the different fibers for composites will be discussed, in light of future potentials in engineering applications. By bringing the disciplines of wood and plant fiber science closer together, one could hope for a mutual transfer of knowledge, as the two research fields have evolved rather independently and have thus reached different levels of understanding with regard to various aspects, such as characterization methods, fiber treatment, fiber preform processing, and composite manufacturing.

The paper presents an introduction to (i) the morphology, chemistry and ultrastructure of the fibers, (ii) the modeling of the mechanical properties of the fibers, (iii) the fiber preforms available for manufacturing of composites, (iv) the typical mechanical properties of the composites, (v) the modeling of the mechanical properties of the composites with special focus on composites having a random fiber orientation and a non-negligible porosity content, and (vi) the moisture sensitivity of the composites. Furthermore, examples of new composite applications are given, followed by considerations of the future perspective of using wood and plant fibers to produce cellulose nanofiber composites. Finally, an outline of the differences and resulting advantages of the two types of cellulose fibers, namely, wood and plant fibers, is given, as well as the differences and resulting advantages of cellulose fibers versus synthetic fibers.

MORPHOLOGY, CHEMISTRY, AND ULTRASTRUCTURE OF FIBERS

Plantae is the one of the five kingdoms of living organisms that includes green plants, that is, mosses, ferns, gymnosperms (e.g., softwood), and angiosperms (e.g., hardwood and annual plants). The cells of green plants are surrounded by a rigid cell wall, and this is the main characteristic distinguishing them from cells in animals. In some types of cells, the cell walls are enlarged to have superior mechanical properties, which provide the required structural performance of the plants. The dimensions of these so-called fibers vary between different plants but their overall shape is most often elongated with lengths in the range 1–50 mm, and diameters in the range 15–30 μm. In the perspective of composite reinforcement, it is convenient to group the fibers by their lengths.

- Short fibers (1–5 mm), originating typically from wood species (e.g., spruce, pine, birch, eucalyptus), and typically used for making composites with in-plane isotropic properties, that is, composites with a non-specific (random) fiber orientation.
- Long fibers (5–50 mm), originating typically from annual plant species (e.g., flax, hemp, jute), and typically used for making composites with anisotropic properties, that is, composites with a specific fiber orientation.

In the living green plants, when the fibers are fully developed, their intracellular organelles start to degenerate resulting in fibers having an empty central cavity, the so-called lumen. In wood fibers, the luminal area is in the range 20–70% of the fiber cross-sectional area [9]. In contrast, annual plant fibers, such as hemp and flax, have a relatively smaller luminal area in the range 0–5% [3, 10].

The main chemical constituent of the cell wall is cellulose, which is a non-branched polysaccharide polymer made up of glucose units. For wood fibers, the cellulose chain is having an average length of 5 μm corresponding to a degree of polymerization (i.e., glucose units) of 10,000 [9]. This molecular linearity makes cellulose highly anisotropic with a theoretical stiffness and strength of about 130 and 15 GPa, respectively, in the chain direction [11]. The cellulose chains are arranged in parallel to form bundles, which are denotedmicrofibrils. In some regions of the microfibrils, the glucose molecules of the cellulose chains are arranged in a highly ordered crystalline structure. The two other principal chemical constituents of the cell wall are hemicellulose and lignin. Hemicellulose is a heterogeneous group of polysaccharides characterized by being short and branched. Lignin is a highly branched polymer composed of phenylpropane units organized in a complex three-dimensional structure. In addition to the organization of the three chemical constituents, the structural complexity of the cell wall is increased by being organized into a number of layers differing by the angle of the cellulose microfibrils to the longitudinal fiber axis. The angle of the cellulose microfibrils in the various layers, in addition to the relative layer thicknesses, dictates the overall mechanical performance of the fibers. Thus, altogether, the cell wall of wood and plant fibers is essentially organized like a composite laminate with a number of laminae with differently oriented, stiff and strong semicrystalline cellulose microfibrils embedded in a matrix of hemicellulose and lignin.

In contrast to cellulose fibers, the synthetic fibers that traditionally are used for reinforcement in composites, such as glass and carbon fibers, are monolithic and with a much more simple ultrastructure. Glass fibers are primarily composed of silicon oxide molecules organized in an amorphous configuration. Carbon fibers are composed of carbon atoms in graphite layers that are organized in a stackwise turbostratic configuration.

Table 1 shows key numbers of chemical composition and ultrastructure of cellulose fibers. The cellulose content of unprocessed fibers is in the range of 40–50% w/w for wood fibers, and in the range of 60–70% w/w for plant fibers. Accordingly, the content of hemicellulose and lignin is higher in wood fibers, and this is particularly true for lignin which shows a content of about 30% w/w in wood fibers, in comparison to only about 5% w/w in plant fibers. The chemical composition of wood and plant fibers is clearly different from each other. In addition, wood fibers show lower cellulose crystallinity than plant fibers, with typical values in the ranges of 55–70 and 90–95% w/w, respectively. The microfibril angle in wood fibers vary in the range 3–50° depending on the type and location of the fibers in the wood (e.g., late and early wood) [12], whereas the microfibril angle in plant fibers is more constant in the range 6–10° [13].

Table 1: Chemical composition and ultrastructure of wood and plant fibers

	Chemical composition			Ultrastructure		
	Cellulose (% w/w)	Hemicellulos (% w/w)	Lignin (% w/w)	Microfibril angle (degrees)	Cellulose crystallinity (% w/w)	Reference

Wood fibers						
Spruce	49	20	29		67	[25]
Pine	42	29	28		57	[26]
Pine(kraft pulp)	76				68	[27]
Cedar	44	21	30		57	[28]
Balsa	48	28	22		56	[26]
Birch	41	32	22		54	[26]
Poplar	39	28	30		54	[28]
Soft wood				3–50		[12]
Plant fibers						
Hemp	63	10	6		96	[25]
Hemp	64	14	3		94	[29]
Hemp (retted)	74	12	5		92	[20]
Hemp (scutched)	66	15	5		98	[29]
Flax (cottonized)	76	14	2		88	[29]
Hemp (textile)	91	7	2			[14]
Hemp and flax				6–10		[13]

The effect of processing treatments on the chemical composition of the fibers is shown in Table 1. In general, for both wood and plant fibers, the cellulose content is increased after processing treatments, due to removal of non-cellulose residues of the fibers (e.g., pectins and waxes). The effect of processing treatment is most clearly seen for the highly processed textile hemp fibers in the study by Madsen et al. [14] where the cellulose content was measured to be as high as 91% w/w.

The influence of growth conditions and processing treatments on the chemistry and ultrastructure of cellulose fibers lead typically to fibers with more variable properties than seen for synthetic fibers. This is frequently considered to be one of the major disadvantages of using cellulose fibers for reinforcement in composites. It is however believed that this concern is caused by a general uncertainty about the cause for the variability in properties, and the lack of a system for classification of the quality of cellulose fibers, for example, similar to the system that exist for classification of solid wood. It should also be mentioned that variability in mechanical properties of fibers can have a positive effect on the notch sensitivity and the fracture toughness of composites [15, 16].

Wood Fiber Preforms

Wood fibers are available at a low cost as pulp fibers (Figure 2(a)). These are used to make paper sheets or board materials for packaging. One way to make composites based on wood fibers is to use such fiber mats (Figure 2(b)), which can be impregnated by using for example, a resin transfer molding technique (e.g., [46]). A viscous thermoset resin is impregnating the enclosed wood fiber mat by the aid of a pressure vessel attached to the mold inlet and sometimes also assisted by vacuum suction at the outlet. This manufacturing technique is only adequate for low-viscosity resins, typically thermosets. Thermoplastics usually have a high viscosity in the molten state, and resin transfer molding is not suitable since the impregnation times would be too high, or the required high pressure would induce severe deformation of the fiber mat. Instead, a commingling technology can be advocated. By using papermaking machines, for example, so-called French or Finnish sheet formers in the laboratory scale, one can produce mats composed of commingled wood pulp fibers and thermoplastic fibers (e.g., [46]). The thermoplastic fibers should preferably have similar dimensions as the pulp fibers, in order to have approximately the same hydrodynamic properties during the formation process, which facilitates efficient mixing. Thermoplastic fibers can be spun to have diameters around 30 μm and be chopped to roughly 3 mm lengths (similar to the dimensions of the pulp fibers). When the commingled fiber mat has been dried, it can be placed in a hot press and composite components can be molded. This method is not only limited to flat plates for materials testing, but complex parts with double curvatures can also be made [47].

(a)

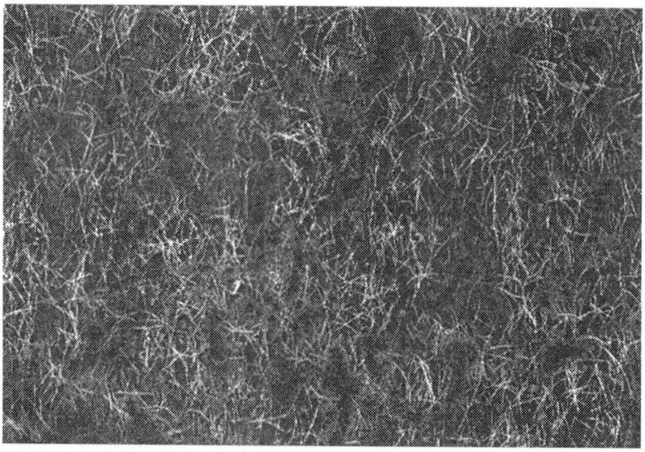

(b)

Figure 2: Wood fiber preforms: (a) pulp made from bleached softwood fibers (courtesy of Innventia, Sweden), and (b) wood fiber mat showing fiber distribution (image height ~ 10 mm).

The papermaking industry encompasses a huge infrastructure to produce wood fiber mats. Anticipated volumes of such fiber preforms for composite applications are extremely small compared with produced volumes of conventional paper and board. Nevertheless, there is an opportunity to build upon the experiences and use small-scale paper mills to produce composite preforms. In the laboratory scale, two main techniques are used to mimic the paper manufacturing process. The most common are sheets produced by dynamic sheet forming and regular handsheets. In dynamic sheet forming, a fiber suspension jet is directed towards a rotating wire drum [48]. The fibers will deposit onto the wire whereas the water goes through the wire. Depending on the jet to wire speed difference, the fibers will orient along the machine direction (circumferential direction of the rotating drum). An anisotropic sheet can then be obtained, with similar features as those manufactured in paper mills, where the fibers are preferably oriented in the machine direction compared with the cross direction. In the handsheets, the fibers are mixed with water in a large container. The water is abruptly let out at the bottom of the container, and the fibers are deposited on a flat wire at the bottom. The fibers are then predominantly randomly oriented in the plane. Thus, the main difference between sheets that are formed using a dynamic sheet former and handsheets is that the former are generally in-plane anisotropic, whereas the latter are in-plane isotropic.

Plant Fiber Preforms

The types of plant fiber preforms available for composites are shown in Figure 3. Here follows descriptions of their processing and characteristics.

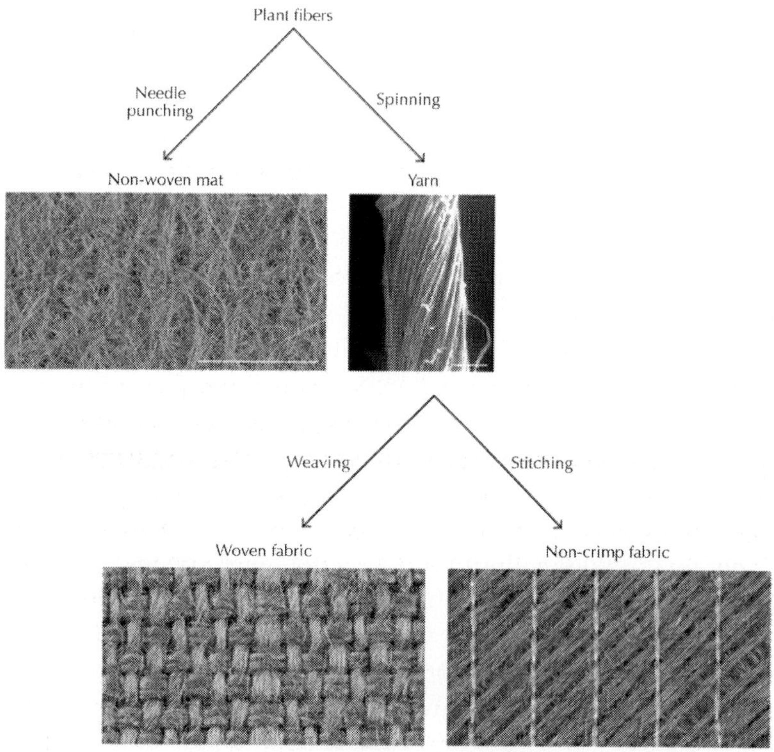

Figure 3: Plant fiber preforms. Scale bar is 10 mm for the mat and fabrics, and 0.1 mm for the yarn.

After the fibers have been extracted from the plants by a retting process, followed by a series of mechanical processes, the fibers can be converted into non-woven mats by air-laid and needle-punching techniques [49]. The fiber orientation in non-woven mats is nominally in-plane random, but they can show a preferred fiber orientation in the machine direction [50]. Alternatively, the fibers can be converted into a continuous yarn by using various spinning techniques, such as ring spinning, rotor spinning, wrap spinning, and air-jet spinning [51]. Ring spinning is the most widely used method. During spinning, the continuous bundle of almost parallelized fibers (a so-called sliver) is twisted so that the fibers take up a helical configuration. The effect of the fiber twisting angle on the mechanical properties of composites has been addressed in a few studies [52–54]. Furthermore, the cross-sectional area of the yarn (which is specified indirectly by its linear density given in units of g/1000 m), and the degree of yarn compaction are other important yarn characteristics, which however have received limited attention in the perspective of composite reinforcement [14]. It can be speculated that the degree of yarn compaction is correlated with the permeability of the yarn for matrix impregnation during manufacturing of composites. More studies are needed to improve the understanding of the correlation between the various structural characteristics of plant fiber yarns, and the mechanical performance of the yarns in composites.

Plant fiber yarn preforms can be used directly to produce composites by commingled filament-winding together with a thermoplastic filament yarn, followed by compression molding [41], or the yarn can be used to make preforms of woven fabrics and non-crimp fabrics. Woven fabrics are fabricated with a range of weaving patterns, such as plain, twill and satin weave, in which the yarns are differently interlaced in the two main, orthogonal, planar directions. The yarns in the two directions can have different linear densities, and they can be placed with different distances to each other. The woven fabrics offer the possibility of having a planar yarn configuration in two dimensions designed to meet the loading profile of a given composite application. Woven fabrics of flax, jute and cotton fibers are widely available, but they are most often tailored for textile applications, and not for composite applications. Non-crimp fabrics consist of yarns that are not held together by being woven into each other, but instead they are stitched together by thin and flexible threads (typically thermoplastic polyester). This means that the yarns are fully stretched; that is, they have no crimp, since they do not have to go over and under each other. Single layers of parallel yarns held together by transversely directed stitching threads are denoted uniaxial non-crimp fabrics. Such uniaxial layers are stacked and stitched together to form biaxial or multiaxial non-crimp fabrics with specific planar yarn orientations, for example, ±45°, 0°/90°, and 0°/+45°/−45°/90°. Recently, a number of European companies have started production of non-crimp fabrics of flax fibers. Thus, for the first time, fabrics of plant fibers that are specifically tailored for composites are commercially available.

MECHANICAL PROPERTIES OF COMPOSITES

The mechanical properties of wood and plant fiber composites have been extensively characterized and analyzed. However, mostly tensile properties, as well as bending and to some extent also impact properties have been characterized, since they are relatively straightforward to measure, and they are commonly used to benchmark different materials in the process of materials development. Other more complex mechanical properties, such as fatigue [55–58] and creep [59] have been studied to a lesser extent.

Table 2 presents typically reported tensile properties (stiffness and strength) of wood and plant fiber composites, together with values for glass and carbon fiber composites. The remarkably high stiffness and strength on 26 GPa and 247 MPa, respectively, for Kraft paper impregnated phenol formaldehyde composites [34] have hitherto not been reached for wood fiber composites. These materials were developed during World War II for use in skins of aircraft wings. Apart from these extreme results by Cox and Pepper [34], it can be observed that cellulose fiber composites (both wood and plant fibers) with a nominal in-plane random fiber orientation, made by using the preforms of loose fibers, paper, and non-woven mats, possess moderate tensile properties with stiffnesses in the range 4–8 GPa and strengths in the range 30–60 MPa. With respect to glass fiber composites, with a similar in-plane random fiber orientation, showing stiffnesses in the range 5–7 GPa and strengths in the range 80–100 MPa, cellulose fiber composites show in general comparable stiffnesses, and slightly lower strengths. It is well known that various chemical approaches can be used to control the interface bonding in order to improve the strength of cellulose fiber composites. Acetylation is one type of surface treatment that can be used to reduce the polarity of the fibers making them more compatible with the (typically) non-polar matrix [60]. Also, coupling agents, such as maleic anhydride, can be used to form covalent bonds between the fibers and the matrix [61]. In the study by Clemons [30] (Table 2), the strength of wood fiber/PP composites were shown to increase from 28 to 52 MPa by using maleic anhydride as a coupling agent, whereas the stiffness did not show any changes. A similar large increase in strength from 40 to 60 MPa has been found in the study of jute fiber/PP composites by Andersen and Plackett [37].

Table 2: Tensile properties of wood and plant fiber composites the type of fiber preforms used for the composites are given, in addition to their nominal fiber orientation; in-plane random (RD) and unidirectional (UD). For means of comparison, tensile properties of glass and carbon fiber composites are shown

	Fiber content (% v/v)	Tensile properties		
		Stiffness (GPa)	Strength (MPa)	Reference
Wood fiber composites				
Wood pulp/PP1; RD	27	4.2	28	[30]
Eucalyptus saw dust/UP1; RD	46	6.2	60	[31]
Kraft + TMP/PP; RD	40	4.5	43	[32]
Sulphite pulp/PP1; RD	50	3.9	51	[33]
Kraft/PF—paper; RD	72	a26.2	247	[34]
Kraft/PF—paper; RD	72	b11.7	156	[34]
Plant fiber composites				
Flax/starch—loose fibers1; RD	37	8.3	51	[35]
Jute/PP—non-woven mat; RD	32	8.4	39	[36]
Jute/PP—non-woven mat; RD	30	5.2	40	[37]
Flax/PLA—non-crimp fabric; UD	39	19.5	150	[38]

Flax/epoxy—non-crimp fabric; UD	35	19.8	234	[39]
Flax/epoxy—yarn2; UD	40	28.0	133	[40]
Flax/PET—yarn2; UD	48	32.0	344	[41]
Glass fiber composites				
Glass/PP—loose fibers1; RD	30	7.3	100	[42]
Glass/PP—chopped strand mat; RD	20	5.4	77	[44]
Glass/epoxy—roving; UD	55	39.0	1080	[45]
Glass/PP—roving2; UD	60	45.0	1020	[43]
Carbon fiber composites				
Carbonc/epoxy—roving; UD	60	313.0	1140	Calculated
Carbond/epoxy—roving; UD	60	142.0	2140	Calculated

[1]Injection molding, [2]filament-winding.

[a]Machine direction; [b]cross direction; [c]high modulus fibers; [d]high strength fibers.

When preforms with nominally unidirectional fibers, such as yarns and non-crimp fabrics are used, the tensile properties of the composites are markedly increased with stiffnesses in the range 20–32 GPa and strengths in the range 130–340 MPa (Table 2). In addition, in comparison to the composites with an in-plane random fiber orientation, the fiber volume content of the unidirectional composites is in general higher (up to 50–55%) [62]. This is related to the better fiber packing ability of aligned fibers as compared to randomly oriented fibers. However, in comparison to synthetic fibers, assemblies of cellulose fibers generally have a lower packing ability [63], which means that the maximum fiber volume content is typically lower in cellulose fiber composites. This is part of the explanation for the lower stiffness of unidirectional cellulose fiber composites, as compared to glass fiber composites, with values of about 30 and 45 GPa, respectively. The contributing stiffness of cellulose fibers in composites has been estimated to be in the range of 20–90 GPa [41], which for the best quality cellulose fibers is comparable to glass fibers with stiffnesses in the range 70–87 GPa [45]. In terms of strength, unidirectional cellulose fiber composites show radically lower values of about 300 MPa compared to about 1000 MPa for glass fiber composites. The explanation for the low strength of unidirectional cellulose fiber composites is currently not known, however, it is expected that fiber defects, which are introduced to the fibers during their processing, play a large role [64].

The tensile properties in Table 2 well illustrate the current status of cellulose fiber composites where stiffness is acceptable, and comparable to glass fiber composites, but strength needs to be improved. Due to the low density of cellulose fibers, the specific mechanical properties of cellulose fiber composites is particularly competitive compared with glass fiber composites Furthermore, if these specific properties are normalized with respect to cost, cellulose fiber composites compare well also with carbon fiber composites. In other words, for large volume applications where weight is an issue, for example, in packaging and transport, cellulose fiber composites are likely to be the main contending materials.

MODELING OF MECHANICAL PROPERTIES OF COMPOSITES

For composite materials, the quantitative relation between microstructure and mechanical properties is generally termed micromechanics, and it has been the scope of extensive research for high-performance composites. Micromechanical models developed for these materials are generally applicable also for cellulose fiber composites, with some modifications to account for the specificities of cellulose fibers. By far, the relation between microstructure and elastic properties is the one that has attracted most attention. Stiffness is one of the foremost design parameters, and it is also amenable to modeling efforts since stiffness represents an average global property, unlike strength which is typically controlled by the locally largest defects in the materials.

Composites with Random Fiber Orientation

Composites with an in-plane random fiber orientation distribution, which is usually the case for wood fiber composites, can be regarded as a stack of unidirectional plies, where the relative thickness of each ply is determined from the fiber orientation distribution. This is known as a laminate analogy, where classic laminate mechanics can be used to relate the elastic properties of the hypothetical unidirectional ply to those of the composite plate. The laminate analogy is schematically illustrated in Figure 4. It is assumed that the fiber orientation distribution is symmetric, which is typically the case for wood fiber mats produced with conventional wet-forming techniques [65]. This means that the materials are globally orthotropic, and the global stiffness matrix can be described by five elastic constants, namely the longitudinal and transverse Young's moduli, the major and minor Poisson ratios, and the shear modulus. The components in the global stiffness matrix can be determined from standardized macroscopic testing, and the fiber orientation distribution can be found by image analysis of scanned sections of the fiber mat [66].

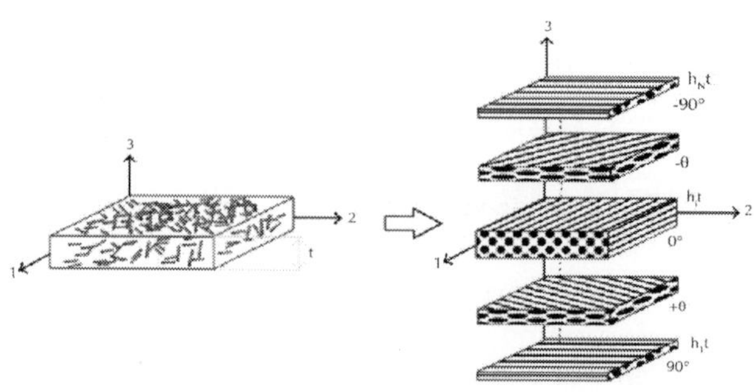

Figure 4: Schematic illustration of a laminate analogy for randomly oriented cellulose fiber composites.

Softwood pulp fibers have an aspect ratio of about 100 [67]. From a mechanical point of view, these fibers can be regarded as continuous, that is, of infinite length, since the ineffective lengths close to the fiber ends are relatively small, as can be calculated by shear-lag theories [68, 69]. The stiffness contribution of wood fibers to the unidirectional plies in the laminate analogy can then be described by simple mechanical models, such as the rule of mixtures model for the longitudinal elastic properties, and the Halpin-Tsai model for the transverse and shear elastic properties (e.g., [70]). For the off-axis properties, Hashin's concentric cylinder model is more accurate [71]. The latter model has been used by Neagu et al. [72] to back-calculate the contributing stiffness of wood fibers from the measured stiffness of composites, and thereby ranking different chemical treatments of wood fibers with respect to their reinforcement efficiency in composites. An example is shown in Figure 5, where the effect of bleaching and retained lignin (characterized by the kappa number) on the contributing fiber stiffness is plotted. This serves as an illustration on how the micromechanical approach can be used to find the optimal bleaching level irrespective of the fiber content and fiber orientation in the composites. These last two parameters may be hard to control in a reproducible manner in the manufacturing of composites.

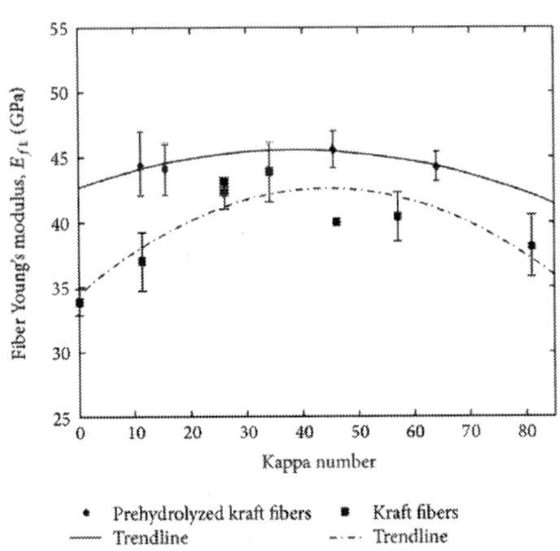

Figure 5: The effect of bleaching level, as characterized by the kappa number, on the contributing fiber stiffness in softwood fiber composites. Shown are results from laboratory softwood kraft fibers and prehydrolyzed kraft fibers [72].

Composites with Non-Negligible Porosity Content

In cellulose fiber composites, the porosity typically makes a noteworthy contribution to the overall composite volume with porosity contents up to 10% [62]. In contrast, in glass and carbon fiber composites, considerable knowledge has been accumulated to diminish the porosity contents below 1% [73]. Altogether, porosity can typically not be neglected in cellulose fiber composites, and it should be integrated in the evaluation of composite performance.

Figure 6 shows examples of the three types of porosity that typically can be found in cellulose fiber composites: fiber lumen porosity, interface porosity, and impregnation porosity. In a study by Madsen et al. [62], the porosity content is correlated with the fiber and matrix contents, and a model for the numerical correlation between weight and volume contents of the composite constituents is presented. Input parameters are (i) the density of fibers and matrix, which can be measured by pycnometry and buoyancy methods, (ii) a number of empirical porosity constants, which can be measured from images of composite microstructures, and (iii) the maximum obtainable fiber volume fraction, which can be determined from the compaction behavior of the fiber assembly. The model predicts the volume fractions of fibers, matrix and porosity as a function of the fiber weight fraction. The model applies to composites in general, but it is particular relevant to composites with a relatively high porosity content, which is typically the case for cellulose fiber composites.

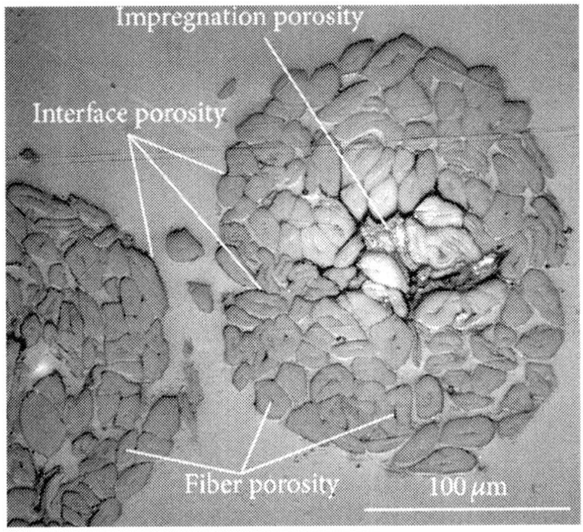

(a)

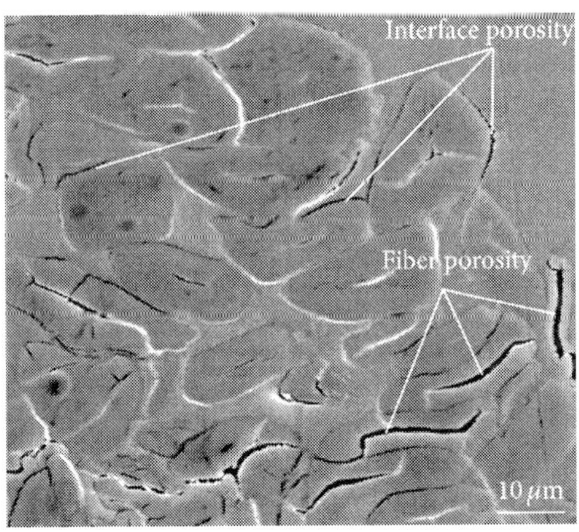

(b)

Figure 6: Types of porosity in cellulose fiber composites. Shown are cross-sectional images of unidirectional hemp fiber/polyethylene terephthalate composites. The optical microscope image in (a) shows a hemp fiber yarn and the scanning electron microscope image in (b) shows a close-up of the fibers in the yarn [62].

Figure 7(a) shows experimental data and model predictions of the volumetric composition of a series of unidirectional flax fiber/thermoplastic matrix composites with variable fiber weight fractions. The volume fractions of fibers and porosity are increased as a function of the fiber weight fraction, until a certain value where after the fiber volume fraction is constant, and the porosity starts to increase more dramatically. The transition fiber weight fraction is determined to be 0.61. Thus, the given composites should be manufactured with a fiber weight fraction of 0.61 to have the best possible combination of high fiber volume fraction, and low porosity, and as will be shown next, this leads to composites with a maximum obtainable tensile stiffness.

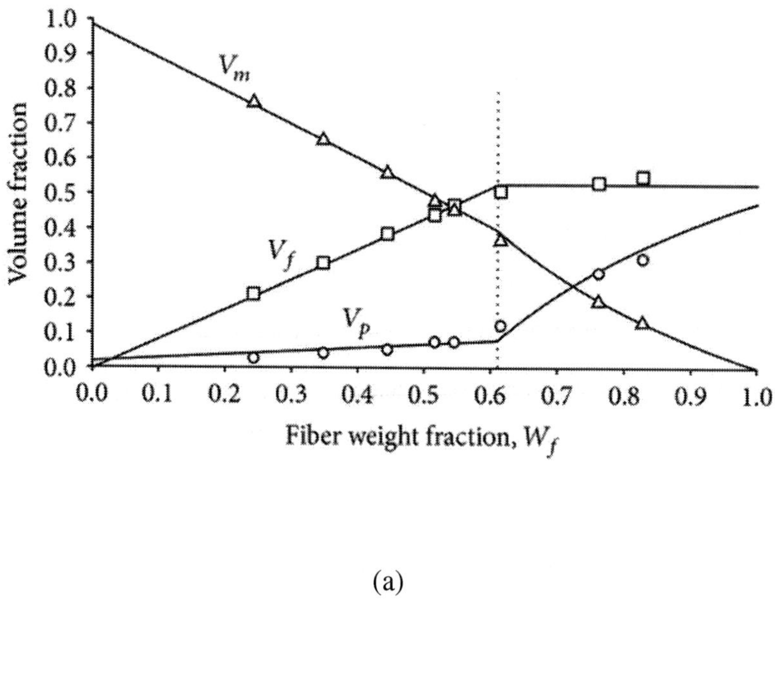

(a)

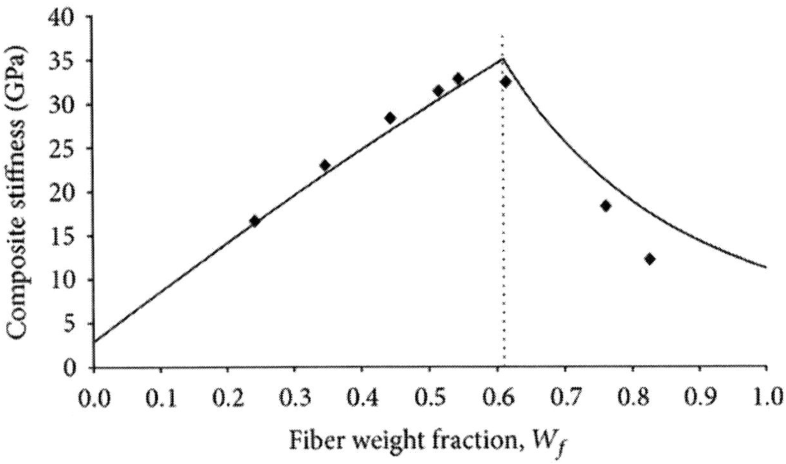

(b)

Figure 7: Experimental data and model predictions of (a) volumetric composition (V_f, V_m and V_p) and (b) stiffness as a function of the fiber weight fraction of unidirectional flax fiber/polyethylene terephthalate composites. Vertical dotted lines indicate the transition fiber weight fraction Data from [75].

The predictions of the volumetric composition in composites can be integrated with micromechanical models. This has been done in the study by Madsen et al. [74] by applying the rule of mixtures model for stiffness of composites. A modified version of the model was used in which the effect of porosity giving stress concentrations in the composites was included. Figure 7(b) shows experimental data and model predictions for stiffness of the unidirectional flax fiber/thermoplastic matrix composites. Stiffness is increased monotonically as a function of the fiber weight fraction until a certain value where after it starts to decrease. The transition fiber weight fraction is 0.61 (i.e., the value determined from the modeling of the volumetric composition of the composites), and here the composites show a maximum obtainable stiffness of about 35 GPa. It demonstrates that the models can be used as guidelines for design of composites with non-negligible porosity content, such as cellulose fiber composites, to have optimal volumetric composition leading to optimal mechanical performance.

MOISTURE SENSITIVITY OF COMPOSITES

Compared to composites with conventional fibers, the Achilles' heel of cellulose fiber composites is their propensity to take up moisture, which leads to swelling, dimensional instability, and potential degradation of mechanical properties. The hydrophilicity of the fibers is due to the abundance of available hydroxyl groups in hemicellulose, in amorphous cellulose and at the surface of cellulose crystallites. For structural materials, moisture sensitivity is generally considered to be a disadvantage, and should be reduced, if possible. For cellulose fiber composites, this can be done by cross-linking of the cell wall polymers in the fibers [76], use of a stiff and hydrophobic matrix [77], and use of a moisture barrier coating [78].

How the swelling of the fibers affects the dimensional stability of the composites is complicated due to the irregular microstructure of the fiber assembly. One way to isolate the hygroexpansion of the fibers, and to quantify its contribution to the hygroexpansion of the composites, is to use micromechanical models. These are similar to models primarily developed for thermal expansion and residual stresses in ceramic-matrix composites. Thermal expansion and hygroexpansion are governed by the same physical equations, where thermal and hygral strains are governed by temperature and moisture, respectively. The micromechanical models for hygroexpansion of composites include also parameters for the elastic properties of the fiber and matrix constituents. In a study by Neagu et al. [48], curl measurements of strips of wood fiber composites and wood fiber mats were used to determine the transverse hygroexpansion coefficient of wood fibers to be approximately 0.10 strains per relative moisture content. This is in accordance with a few scarce data found in the literature on the hygroexpansion properties of cellulose fibers [79].

In a study by Madsen et al. [80] of the hygroexpansion of unidirectional hemp fiber/thermoplastic matrix composites, the dimensional changes were found to be well correlated with the fiber volume fraction. Selected results are shown in Figure 8. By using a reference humidity of 65% RH, the transverse hygroexpansion was found to be 0.9% and 3.3% at humidities of 85 and 100% RH, respectively, for composites with the highest fiber volume fraction of 0.43. The hygroexpansion in the longitudinal direction was found to be low, and slightly negative, which presumable is due to moisture induced relaxation of residual tensile stresses in the matrix. It can be observed in the figure that the experimental data points are well simulated by the micromechanical model lines.

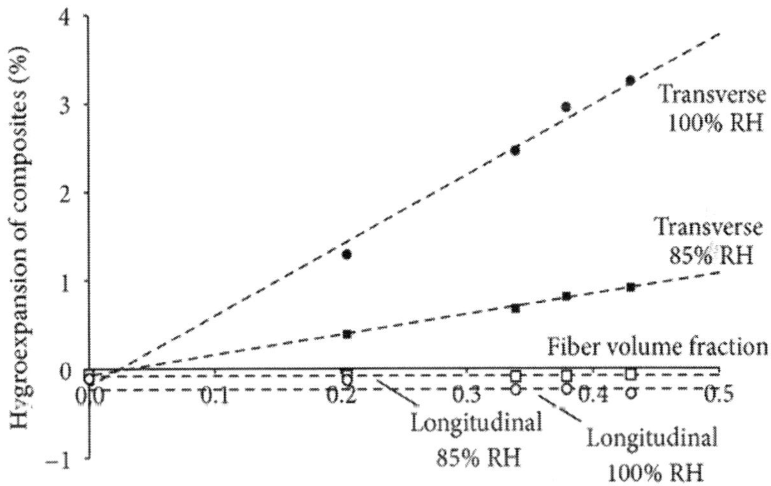

Figure 8: Hygroexpansion of unidirectional hemp fiber/polyethylene terephthalate composites as function of the fiber volume fraction. Hygroexpansion was measured in the longitudinal and transverse directions, and at the humidities of 85 and 100% RH, with respect to a reference humidity of 65% RH. Lines are calculated by micromechanical models Modified from [80].

Glass and carbon fibers do not take up any moisture, although glass fibers are sensitive to environmental stress corrosion in the presence of moisture and tensile stress [81]. The moisture sensitivity is larger for wood fibers than for plant fibers, since the former contains a larger relative amount of hemicellulose which is the most hydrophilic polymer in the cell wall.

APPLICATIONS OF COMPOSITES

In Europe, cellulose fiber composites are mainly used by the automotive industry. The applied fiber preforms are loose fibers used for injection molding techniques, and non-woven mats used for compression molding techniques. Due to the nominal random fiber orientation in these composites, they possess only moderate mechanical properties (see Table 2), but this makes them nevertheless well qualified to be used in non-structural components such as door liners, boot liners, and parcel shelves. The low prices of loose fibers and non-woven mats of cellulose fibers, compared to their synthetic counterparts, form a strong motivation for the use of these two preforms in the automotive industry. Outside Europe, the use of non-structural components based on cellulose fibers is more widespread, and wood fibers are by far the preferred fiber type. In North America, the main applications are building components, such as deckings, windows profiles and floorings.

Recently, in the context of research and development projects, a number of demonstrators have been made to reveal the good potential of cellulose fiber composites in new kinds of applications (see Figure 9).

- Sculpture shown at the Louisiana Museum of Modern Arts, Denmark, special exhibition "Green architecture for the future", winner of the JEC innovation award 2010, joint venture of 20 companies coordinated by 3XN architects, Denmark.
- Wheel rim, EU 7th Framework Programme project, NATEX (2008–2012).
- Chair for children, developed by the research institute Innventia AB together with pulp industry and architects, exhibited at the Milan furniture fair 2009.
- Double-curvature panels designed by M. Larsen and K.R. Nielsen, exhibited at "Klimaforum09/Ideas at work" in connection with COP15, Copenhagen 2009, and at the JEC exhibition 2010.
- Small-scale rotor blade to be used for a wind turbine car, EU 7th Framework Programme project, WOODY (2009–2012).
- Exhibition stands at the Swedish Nautical Historical Museum in Stockholm, Sweden, EU 7th Framework Programme Project, WOODY (2009–2012).

Figure 9: Examples of new applications of cellulose fiber composites: (top, left) sculpture, (top, right) wheel rim, (middle, left) chair for children, (middle, right) double-curvature panels, (bottom, left) small-scale rotor blade, and (bottom, right) exhibition stand. See text for more details.

FUTURE PERSPECTIVES: NANOFIBER COMPOSITES

In recent years, considerable attention has been directed towards composites made from cellulose nanofibers (e.g., [82]). As already described, the cell wall of wood and plant fibers is structured like composites with cellulose microfibrils embedded in a matrix of hemicellulose and lignin. The cellulose microfibrils are having lateral dimensions in the 10–100 nm range and axial dimensions in the micrometer range, and they are therefore suitable as reinforcement in nanofiber composites. The idea is to achieve considerable improvements in engineering properties with the addition of nanofibers, beyond those obtained with fibers in the micrometer range. This can be attributed to the high specific surface area of the nanoscale fibers, which will affect the properties of the surrounding matrix. The success of nanofiber composites is particularly obvious if only a minute addition of fibers is considered, and the dispersion of nanofibers is preserved.

Carbon nanotube composites have shown great promise for a relatively long time, but have yet to deliver in large volume applications [83]. A difference between cellulose nanofibers and carbon nanotubes is the ability of the cellulose nanofibers to bond to each other, by hydrogen bonding, whereas the carbon nanotube surface is chemically inert. The cellulose nanofibers can form a very strong network, and furthermore bond well to polymer matrix materials with polar groups. This leads however also to processing difficulties, since the cellulose nanofibers tend to aggregate and take a long time to dry after wet processing. Processability and performance are thus complementary and mutually opposing behaviors. The functional hydroxyl groups of the fibers can be modified to improve dispersion and processability, although this is typically accompanied with increased costs. The main challenges for cellulose nanofiber composites are probably to learn how to manufacture bulk composite components with retained nanofiber slenderness and dispersion. The raw materials are the cellulose fibers themselves, from wood pulp or plant fibers, making the raw materials costs negligible compared with manufacturing costs.

OVERALL COMPARISON BETWEEN FIBERS

As shown in the sections above, wood and plant fibers are similar in some respects and differ in others. Thus, depending on the intended application, one particular fiber type is more suitable than the other. In the following, an overall comparison is given to highlight some advantages of wood versus plant fibers, and vice versa. Similarly, cellulose fibers are compared to their synthetic counterparts, glass and carbon fibers.

Advantages of wood fibers, as compared with plant fibers, are as follows.

- Low cost, readily available from pulp mills.
- Relatively short fibers mean better processability.
- Mature infrastructure available in pulp and paper mills to produce large quantities at low cost.
- Preforms can be made using paper-making technologies.
- Rather uniform batches of pulp qualities can be achieved.
- Does not compete with cultivation of food crops.

Advantages of plant fibers, as compared with wood fibers, are as follows.

- High productivity and yield.
- High cellulose content, high degree of cellulose crystallinity, low microfibril angle, small lumen mean excellent mechanical properties of fibers.
- Relatively long fibers mean possibility to control fiber orientation and lay-up.
- Textile technologies can be used to produce yarns, woven fabrics and non-crimp fabrics.

Despite the differences, wood and plant fibers have more in common than in what differ them from one another. Some advantages of cellulose fibers as compared with glass and carbon fibers can be mentioned.

- Renewable.
- Biodegradable.
- Light, that is, the composites have good specific properties which are important in automotives and packaging.
- Low cost raw materials.

The main disadvantages of cellulose fibers as compared with glass and carbon fibers are as follows.

- Moderate mechanical properties.
- Sensitivity to moisture, leading to dimensional instability, and potential degradation of mechanical properties.
- Not fully developed composite manufacturing techniques.

The above lists are by no means comprehensive, but only serve to show some of the traits of wood and plant fibers in an applied composite context. In the further development of cellulose fiber composites, both advantages and disadvantages play an important role. The specific advantages guide which application areas that are relevant. For instance, the combination of low cost, renewability and biodegradability make cellulose fiber composites suitable materials for packaging applications. The disadvantages limit their applications. Research on how to alleviate these shortcomings can expand the proliferation of cellulose fibers as an eco-friendly alternative to synthetic fibers. If cellulose fibers can be processed to retain better their innate high stiffness and strength, and they can be modified to become less hydrophilic they are also potential reinforcement fibers in advanced structural outdoor applications, for example, in rotor blades for wind turbines and in load-carrying components in transport applications.

REFERENCES

1. A. K. Mohanty, M. Misra, and L. T. Drzal, Natural Fibers, Biopolymers, and Biocomposites, CPR Press, Boca Raton, Fla, USA, 2005.
2. K. L. Pickering, Properties and Performance of Natural-Fibre Composites, Woodhead Publishing Limited, Cambridge, UK, 2008.
3. J. Müssig, Industrial Applications of Natural Fibres—Structure, Properties and Technical Applications, John Wiley & Sons, Chichester, UK, 2010.
4. CELC European Scientific Committee, "Flax and hemp fibres: a natural solution for the composite industry," in European Hemp and Flax Confederation, F. Reux and I. Verpoest, Eds., JEC Composites Group, Paris, France, 2012.
5. C. Hill and M. Hughes, "Natural fibre reinforced composites opportunities and challenges," Journal of Biobased Materials and Bioenergy, vol. 4, no. 2, pp. 148–158, 2010.
6. E. Zini and M. Scandola, "Green composites: an overview," Polymer Composites, vol. 32, pp. 1905–1915, 2011.

7. O. Faruk, A. K. Bledzki, H. Fink, and M. Sain, "Biocomposites reinforced with natural fibers: 2000–2010," Progress in Polymer Science, vol. 37, pp. 1552–1596, 2012.
8. A. Shahzad, "Hemp fiber and its composites—a review," Journal of Composite Materials, vol. 46, pp. 973–986, 2012.
9. J. F. Siau, Wood: Influence of Moisture on Physical Properties, Department of Wood Science and Forest Products, Virginia Polytechnic Institute and State University, Burruss Hall, Va, USA, 1995.
10. D. R. Perry, Identification of Textile Materials, The Textile Institute, Manchester, UK, 7th edition, 1985.
11. H. Lilholt and J. M. Lawther, "Natural organic fibers," in Comprehensive Composite Materials, A. Kelly and C. Zweben, Eds., vol. 1, chapter 10, pp. 303–325, Elsevier Science, 2000.
12. S. E. Anagnost, R. E. Mark, and R. B. Hanna, "Variation of microfibril angle within individual tracheids," Wood and Fiber Science, vol. 34, no. 2, pp. 337–349, 2002.
13. J. Gassan, A. Chate, and A. K. Bledzki, "Calculation of elastic properties of natural fibers," Journal of Materials Science, vol. 36, no. 15, pp. 3715–3720, 2001.
14. B. Madsen, P. Hoffmeyer, A. B. Thomsen, and H. Lilholt, "Hemp yarn reinforced composites I. Yarn characteristics," Composites Part A, vol. 38, no. 10, pp. 2194–2203, 2007.
15. J. Lindhagen and L. Berglund, "Microscopical damage mechanisms in glass fiber reinforced polypropylene," Journal of Applied Polymer Science, vol. 69, no. 7, pp. 1319–1327, 1998.
16. A. Ridruejo, C. Gonzalez, and J. LLorca, "Damage micromechanisms and notch sensitivity of glass-fiber non-woven felts: An Experimental and Numerical Study," Journal of the Mechanics and Physics of Solids, vol. 58, pp. 1628–1645, 2010.
17. R. C. Neagu, E. K. Gamstedt, S. L. Bardage, and M. Lindström, "Ultrastructural features affecting mechanical properties of wood fibers," Wood Materials Science & Engineering, vol. 1, pp. 146–170, 2006.
18. L. Salmén and I. Burgert, "Cell wall features with regard to mechanical performance. A review. COST Action E35 2004–

2008: wood machining—Micromechanics and fracture," Holzforschung, vol. 63, no. 2, pp. 121–129, 2009.

19. L. Salmén, "Micromechanical understanding of the cell-wall structure," Comptes Rendus—Biologies, vol. 327, no. 9-10, pp. 873–880, 2004.
20. A. Thygesen, A. B. Thomsen, G. Daniel, and H. Lilholt, "Comparison of composites made from fungal defibrated hemp with composites of traditional hemp yarn," Industrial Crops and Products, vol. 25, no. 2, pp. 147–159, 2007.
21. K. Hofstetter, C. Hellmich, and J. Eberhardsteiner, "Development and experimental validation of a continuum micromechanics model for the elasticity of wood," European Journal of Mechanics A, vol. 24, no. 6, pp. 1030–1053, 2005.
22. K. Hofstetter, C. Hellmich, and J. Eberhardsteiner, "The influence of the microfibril angle on wood stiffness: a continuum micromechanics approach," Computer Assisted Mechanics and Engineering Sciences, vol. 13, no. 4, pp. 523–536, 2006.
23. D. H. Page, F. El-Hosseiny, and K. Winkler, "Behaviour of single wood fibres under axial tensile strain,"Nature, vol. 229, no. 5282, pp. 252–253, 1971.
24. L. Salmén and A. de Ruvo, "A model for the prediction of fiber elasticity," Wood Fiber Science, vol. 17, pp. 336–350, 1985.
25. A. Thygesen, J. Oddershede, H. Lilholt, A. B. Thomsen, and K. Ståhl, "On the determination of crystallinity and cellulose content in plant fibres," Cellulose, vol. 12, no. 6, pp. 563–576, 2005.
26. E. Sjöström, Wood Chemistry, Fundamentals and Applications, Academic Press, San Diego, Calif, USA, 2nd edition, 1993.
27. T. Liitiä, S. L. Maunu, B. Hortling, T. Tamminen, O. Pekkala, and A. Varhimo, "Cellulose crystallinity and ordering of hemicelluloses in pine and birch pulps as revealed by solid-state NMR spectroscopic methods," Cellulose, vol. 10, no. 4, pp. 307–316, 2003.
28. R. Rowell, the Chemistry of Solid Wood, American Chemical Society, Washington, DC, USA, 1984.
29. A. Thygesen, B. Madsen, A. B. Bjerre, and H. Lilholt, "Cellulosic fibers: effect of processing on fiber bundle strength," Journal of Natural Fibers, vol. 8, pp. 161–175, 2011.

30. C. M. Clemons, "Woodfiber-plastic composites in the United States—history and current and future markets," in Proceedings of the 3rd International Wood and Natural Fiber Composites Symposium, pp. 1–7, Kassel, Germany, September 2000.

31. N. E. Marcovich, M. M. Reboredo, and M. I. Aranguren, "Composites from sawdust and unsaturated polyester," Journal of Applied Polymer Science, vol. 61, no. 1, pp. 119–124, 1996.

32. R. T. Woodhams, G. Thomas, and D. K. Rodgers, "Wood fibers as reinforcing fillers for polyolefins,"Polymer Engineering and Science, vol. 24, no. 15, pp. 1166–1171, 1984.

33. M. Bengtsson, M. L. Baillif, and K. Oksman, "Extrusion and mechanical properties of highly filled cellulose fibre-polypropylene composites," Composites Part A, vol. 38, no. 8, pp. 1922–1931, 2007.

34. H. L. Cox and K. W. Pepper, "Paper-based plastics—part I: the preparation of phenolic laminated boards," Journal of the Society of Chemical Industry, vol. 63, pp. 150–154, 1944.

35. K. Nättinen, S. Hyvärinen, R. Joffe, L. Wallström, and B. Madsen, "Naturally compatible: starch acetate/cellulosic fiber composites. I. processing and properties," Polymer Composites, vol. 31, no. 3, pp. 524–535, 2010.

36. H. Toftegaard, "Tensile testing of jute/PP laminates," Risø Report Risø-I-1824(EN), Risø National Laboratory, Materials Research Department, Roskilde, Denmark, 2002.

37. T. L. Andersen and D. Plackett, Polymer Composite Product, A Process for the Manufacture Thereof and Use of the Product, International Application Number: PCT/DK02/00085. International Publication Number: WO 02/064670 A1, August 2002.

38. B. Madsen, H. Lilholt, A. Thygesen, E. Arnold, B. Weager, and R. Joffe, "Aligned flax fibre/polylactate composites—a materials model system to show the potential of biocomposites in engineering applications," Journal of Nanostructured Polymers and Nanocomposites, vol. 4, no. 4, pp. 139–145, 2009.

39. F. Bottoli and L. Pignatti, Design and processing of structural components in biocomposite materials—Case study: rotor blade for wind turbine cars [M.S. thesis], Materials Research Division,

Risø National Laboratory for Sustainable Energy, Technical University of Denmark, 2011.

40. I. Van de Weyenberg, J. Ivens, A. De Coster, B. Kino, E. Baetens, and I. Verpoest, "Influence of processing and chemical treatment of flax fibres on their composites," Composites Science and Technology, vol. 63, no. 9, pp. 1241–1246, 2003.
41. S. Mehmood and B. Madsen, "Properties and performance of flax yarn/thermoplastic polyester composites," Journal of Reinforced Plastics and Composites, vol. 31, pp. 62–73, 2012.
42. J. L. Thomason, "The influence of fibre length and concentration on the properties of glass fibre reinforced polypropylene. 6. the properties of injection moulded long fibre PP at high fibre content,"Composites Part A, vol. 36, no. 7, pp. 995–1003, 2005.
43. E. K. Gamstedt, L. A. Berglund, andT. Peijs, "Fatigue mechanisms in unidirectional glass-fibre-reinforced polypropylene," Composites Science and Technology, vol. 59, no. 5, pp. 759–768, 1999.
44. K. Oksman, "Mechanical properties of natural fibre mat reinforced thermoplastic," Applied Composite Materials, vol. 7, no. 5-6, pp. 403–414, 2000.
45. A. R. Bunsell and J. Renard, Fundamentals of Fibre Reinforced Composite Materials, Institute of Physics, London, UK, 2005.
46. K. M. Almgren, E. K. Gamstedt, P. Nygård, F. Malmberg, J. Lindblad, and M. Lindström, "Role of fibre-fibre and fibre-matrix adhesion in stress transfer in composites made from resin-impregnated paper sheets," International Journal of Adhesion and Adhesives, vol. 29, no. 5, pp. 551–557, 2009.
47. M. Lindström, F. Berthold, K. Gamstedt, J. Varna, and K. Wickholm, "Hierarchical design as a tool in materials development," in Proceedings of the 10th International Conference on Progress in Biofiber Plastic Composites, pp. 1–7, Toronto, Canada, 2008.
48. R. C. Neagu, E. K. Gamstedt, and M. Lindström, "Influence of wood-fibre hygroexpansion on the dimensional instability of fibre mats and composites," Composites Part A, vol. 36, no. 6, pp. 772–788, 2005.
49. M. Eriksen and B. E. Pallesen, New Generation Airforming for Flax and Hemp, Nonwovens World, 2002.
50. H. S. Kim, B. Pourdeyhimi, A. Abhiraman, and P. Desai, "Characterization of structural changes in nonwoven fabrics

during load-deformation experiments," Journal of Textile and Apparel, Technology and Management, vol. 1, no. 1, pp. 1–6, 2000.

51. P. Grosberg and C. Iype, Yarn Production—Theoretical Aspects, The Textile Institute, Manchester, UK, 1999.
52. S. Goutianos, T. Peijs, B. Nystrom, and M. Skrifvars, "Development of flax fibre based textile reinforcements for composite applications," Applied Composite Materials, vol. 13, no. 4, pp. 199–215, 2006.
53. M. Rask, Microstructure and mechanical properties of aligned natural fibre composites [Ph.D. thesis], Department of Wind Energy, Technical University of Denmark. DTU Wind Energy E, 2012.
54. D. U. Shah, P. J. Schubel, P. Licence, and M. J. Clifford, "Determining the minimum, critical and maximum fibre content for twisted yarn reinforced plant fibre composites," Composites Science and Technology, vol. 72, pp. 1909–1917, 2012.
55. J. Gassan and A. K. Bledzki, "Effect of cyclic moisture absorption desorption on the mechanical properties of silanized jute-epoxy composites," Polymer Composites, vol. 20, no. 4, pp. 604–611, 1999.
56. A. N. Towo and M. P. Ansell, "Fatigue evaluation and dynamic mechanical thermal analysis of sisal fibre-thermosetting resin composites," Composites Science and Technology, vol. 68, no. 3-4, pp. 925–932, 2008.
57. S. Liang, P. B. Gning, and L. Guillaumat, "A comparative study of fatigue behaviour of flax/epoxy and glass/epoxy composites," Composites Science and Technology, vol. 72, pp. 535–543, 2012.
58. D. U. Shah, P. J. Schubel, M. J. Clifford, and P. Licence, "Fatigue life evaluation of aligned plant fibre composites through S-N curves and constant-life diagrams," Composites Science and Technology, vol. 74, pp. 139–149, 2013.
59. J. Gassan and A. K. Bledzki, "Influence of fiber surface treatment on the creep behavior of jute fiber-reinforced polypropylene," Journal of Thermoplastic Composite Materials, vol. 12, no. 5, pp. 388–398, 1999.

60. H. P. S. A. Khalil, H. Ismail, H. D. Rozman, and M. N. Ahmad, "Effect of acetylation on interfacial shear strength between plant fibres and various matrices," European Polymer Journal, vol. 37, no. 5, pp. 1037–1045, 2001.
61. G. E. Myers, I. S. Chahyadi, C. A. Coberly, and D. S. Ermer, "Wood flour/polypropylene composites: influence of maleated polypropylene and process and composition variables on mechanical properties,"International Journal of Polymeric Materials, vol. 15, pp. 21–44, 1991.
62. B. Madsen, A. Thygesen, and H. Lilholt, "Plant fibre composites—porosity and volumetric interaction,"Composites Science and Technology, vol. 67, no. 7-8, pp. 1584–1600, 2007.
63. B. Madsen and H. Lilholt, "Compaction of plant fiber assemblies in relation to composite fabrication," in Proceedings of the 23rd Risø International Symposium on Materials Science. Sustainable Natural and Polymeric Composites—Science and Technology, pp. 239–250, Risø National Laboratory, 2002.
64. M. Hughes, "Defects in natural fibres: their origin, characteristics and implications for natural fibre-reinforced composites," Journal of Materials Science, vol. 47, pp. 599–609, 2012.
65. A. L. Erikkilä, P. Pakarinen, and M. Odell, "Sheet forming studies using layered orientation analysis: can lead to a better understanding of the drainage process," Pulp and Paper Canada, vol. 99, no. 1, pp. 81–85, 1998.
66. F. Gadala-Maria and F. Parsi, "Measurement of fiber orientation in short-fiber composites using digital image processing," Polymer Composites, vol. 14, no. 2, pp. 126–131, 1993.
67. J. Duanmu, E. K. Gamsted, and A. Rosling, "Synthesis and preparation of crosslinked allylglycidyl ether-modified starch-wood fibre composites," Starch, vol. 59, no. 11, pp. 523–532, 2007.
68. H. L. Cox, "The elasticity and strength of paper and other fibrous materials," British Journal of Applied Physics, vol. 3, no. 3, article 302, pp. 72–79, 1952.
69. J. A. Nairn, "On the use of shear-lag methods for analysis of stress transfer in unidirectional composites," Mechanics of Materials, vol. 26, no. 2, pp. 63–80, 1997.

70. E. K. Gamstedt, E. Sjöholm, C. Neagu, F. Berthold, and M. Lindström, "Effects of fiber bleaching and earlywood-latewood fraction on tensile properties of wood-fiber reinforced vinyl ester," in Proceedings of the 23rd Risø International Symposium on Materials Science: Sustainable Natural and Polymeric Composites—Science and Technology, H. Lilholt, Ed., pp. 185–196, Risø National Laboratory, 2002.
71. Z. Hashin, "Stiffness contribution of various wood fibers to composite materials," Journal of Applied Mechanics, vol. 46, no. 3, pp. 543–550, 1979.
72. R. C. Neagu, E. K. Gamstedt, and F. Berthold, "Stiffness contribution of various wood fibers to composite materials," Journal of Composite Materials, vol. 40, no. 8, pp. 663–699, 2006.
73. J. A. E. Månson, M. D. Wakeman, and N. Bernet, "Composite processing and manufacturing—an overview," in Comprehensive Composite Materials, A. Kelly and C. Zweben, Eds., vol. 2, chapter 2, pp. 577–607, Elsevier Science, 2000.
74. B. Madsen, A. Thygesen, and H. Lilholt, "Plant fibre composites—porosity and stiffness," Composites Science and Technology, vol. 69, no. 7-8, pp. 1057–1069, 2009.
75. M. Aslan, S. Mehmood, and B. Madsen, "Effect of consolidation pressure on volumetric composition and stiffness of unidirectional flax fibre composites," Journal of Materials Science, vol. 48, no. 10, pp. 3812–3824, 2013.
76. C. A. S. Hill, H. P. S. A. Khalil, and M. D. Hale, "A study of the potential of acetylation to improve the properties of plant fibres," Industrial Crops and Products, vol. 8, no. 1, pp. 53–63, 1998.
77. K. M. Almgren, E. K. Gamstedt, F. Berthold, and M. Lindström, "Moisture uptake and hygroexpansion of wood fiber composite materials with polylactide and polypropylene matrix materials," Polymer Composites, vol. 30, no. 12, pp. 1809–1816, 2009.
78. C. Esnaashari, S. N. Khorasani, M. Entezam, and S. Khalili, "Mechanical and water absorption properties of sawdust-low density polyethylene nanocomposite," Journal of Applied Polymer Science, vol. 127, pp. 1295–1300, 2013.

79. K. M. Almgren, E. K. Gamstedt, and J. Varna, "Contribution of wood fiber hygroexpansion to moisture induced thickness swelling of composite plates," Polymer Composites, vol. 31, no. 5, pp. 762–771, 2010.

80. B. Madsen, P. Hoffmeyer, and H. Lilholt, "Hemp yarn reinforced composites—III Moisture content and dimensional changes," Composites Part A, vol. 43, pp. 2151–2160, 2012.

81. E. L. Rodriguez, "Corrosion of glass fibres," Journal of Materials Science Letters, vol. 6, no. 6, pp. 718–720, 1987.

82. S. J. Eichhorn, A. Dufresne, M. Aranguren et al., "Review: current international research into cellulose nanofibres and nanocomposites," Journal of Materials Science, vol. 45, no. 1, pp. 1–33, 2010.

83. Y. Dzenis, "Spinning continuous fibers for nanotechnology," Science, vol. 304, no. 5679, pp. 1917–1919, 2004.

Chapter 3

Columnar-to-Equiaxed Transition in Metal-Matrix Composites Reinforced with Silicon Carbide Particles

Alicia E. Ares[1, 2] and Carlos E. Schvezov[1, 2]

[1]Faculty of Sciences, University of Misiones, 1552 Félix de Azara Street, 3300 Posadas, Misiones, Argentina

[2]Member of Scientific Research Career (CIC) of the National Council of Scientific and Technical Research (CONICET), 1917 Rivadavia Street, 1033 Buenos Aires, Argentina

ABSTRACT

The present work is focused on the study of the effect of directional heat extraction on the silicon-carbide (SiC) distribution in zinc-aluminum matrix composites (MMCs) and on the columnar-to-equiaxed (CET) position in directionally solidified samples. To this end, a ZA-27 alloy matrix was reinforced with ceramic particles of SiC and vertically directionally solidified. The cooling rates, temperature gradients, and interphase velocities were then measured, and their influence on the solidification microstructure of the MMCs was analyzed. The recalescence detected and measured during the equiaxed transition was of the order of 3.5°C to 1.1°C. The values of the temperature gradients reached a minimum during the CET and were even negative in most cases (between −3.89 K and 0.06 K). The interphase velocities varied between 0.07 mm/s and 0.44 mm/s at the transition. Also, the presence of ceramic particles in ZA-27 alloys affected the thermodynamic local conditions and the kinetics of nucleation, producing a finer microstructure.

INTRODUCTION

A metal-matrix composite (MMC) is composite material with at least two constituent parts, with one being a metal. The other material may be a different metal or another material, such as a ceramic or organic compound. The matrix is the monolithic material into which the reinforcement is embedded and is completely continuous. The reinforcement does not always serve a purely structural task but is also used to change physical properties such as wear resistance, friction coefficient, or thermal conductivity. The reinforcement can be either continuous or discontinuous [1]. The aspect ratio of the reinforcement is an important quantity, because the degree of load transfer from the matrix to the reinforcement is directly proportional to the reinforcement aspect ratio. Particle or short fiber reinforced metals have a much lower aspect ratio, so they exhibit lower strengths than their continuous fiber counterparts, although the properties of these composites are much more isotropic [1, 2].

Metal-matrix composites can be processed by several techniques. Some of these important techniques are liquid-state processes (casting or liquid infiltration, squeeze casting, or pressure infiltration), solid-state processes (diffusion bonding, deformation processing, powder processing, sinter-forging, and deposition techniques), in situ processes, and spray-forming of particulate MMCs [2].

Metal-matrix particulate composites such as SiC particle-reinforced aluminum can offer a 50–100% increase in Young's modulus over that of unreinforced aluminum, that is, the modulus being equivalent to that of titanium, but density that is about 33% lower. In general, ceramic reinforcements (fibers, whiskers, or particles) have a coefficient of thermal expansion lower than that of most metallic matrices. On the other hand, when the composite is subjected to a temperature change, thermal stresses are generated in both components [3].

The development of composites obtained by solidification of alloys has made remarkable progress and their applications in automotive and aerospace industries have increased in recent decades. Among these applications, the most current one is the zinc and aluminum base.

It is important for metallurgical processes to establish relationships between thermal parameters and solidification structures, because the solidification structure encompasses morphological aspects of different types of grains (columnar, equiaxed, or columnar-to-equiaxed transition) and dendrite structure (primary, secondary, or tertiary) along the distribution and state of aggregation of the second reinforcing phase.

The interaction between the parameters involved in the columnar-to-equiaxed transition (CET) phenomenon, which has gained considerable attention over the last decades, allows understanding the structure of ingot castings and optimizing the industrial practice [4–18].

The characteristics of metal-matrix composite materials are determined by their microstructure and internal interfaces, which are affected by their production and thermal mechanical prehistory. The microstructure covers the structure of the matrix and the reinforced phase. The chemical composition, grain and/or subgrain size, texture, precipitation behavior, and lattice defects are of importance to the matrix. The second phase is characterized by its volume percentage, its kind, its size, its distribution, and its orientation. Local varying internal tension due to the different thermal expansion behavior of the two phases is an additional influencing factor [1, 2, 19].

From the past research it is known that rules developed for solidification of unreinforced metals cannot be applied directly to metal matrix composites. An important tool in the study of alloy solidification is the Bridgman furnace, which produces well-controlled steady-state directional solidification conditions.

Directional solidification can be used to investigate solidification modes, analyzing the thermal parameter (thermal gradient and growth velocity) variation during the solidification process. Nevertheless, alloy solidification in engineering castings is generally dendritic and thermal parameters are neither measured nor constant over the course of solidification [17–21].

The present work is focused on the study of the effect of directional heat extraction on the silicon-carbide (SiC) distribution in an aluminum matrix composite and on the CET position in directionally solidified samples. To this end, Zn-27 wt.% Al (ZA-27) alloy matrix was reinforced with ceramic particulates of SiC and was vertically directionally solidified. The thermal parameters (cooling rates, temperature gradients, and interphase velocities) were then measured, and their influence on the solidification microstructure of the MMCs was analyzed. Experimental results include grain size and particle distribution as a function of the solidification conditions imposed by the metal/mould system. The conditions for the CET in MMCs are compared with those in Zn-Al alloys.

EXPERIMENTAL PROCEDURE

For the preparation of metal-matrix composites (MMCs), zinc (99.98 wt pct) and commercial aluminum (99.96 wt pct) were mixed with SiC particles. Their compositions are shown in Table 1.

Table 1: Liquidus temperature (T_L), solidus temperature (T_S), cooling rate in the liquid ($\dot{T}_L$), cooling rate in the solid ($\dot{T}_S$), liquidus interphase velocity (V_{LC}), minimum CET position ($CET_{MIN.}$), maximum CET position ($CET_{MAX.}$), critical gradients (G_C), and recalescence values (R) obtained from the temperature versus time curves

Experiment number	Alloy and composite	T_L (K)	T_S (K)	$\dot{T}_L$ (K/s)	$\dot{T}_S$ (K/s)	V_{LC} (K/s)	$CET_{MIN.}$ (mm)	$CET_{MAX.}$ (mm)	G_C (K/mm)	R(°C)
1	Zn-27 wt.% Al (ZA27)	778.2	709.7	3.3	1.1	1.05	35	87	−0.09	2.4
2	Zn-27 wt.% Al (ZA27)	778.2	709.7	2.2	0.8	1.4	21	44	0.06	1.1
3	ZA27 + 5vol% SiC	767.8	710.2	2.3	0.9	1.2	26	48	−3.89	2.3
4	ZA27 + 5vol% SiC	767.8	710.2	3.4	1.0	2.4	41	63	−1.52	2.5
5	ZA27 + 8vol% SiC	762.3	706.7	3.1	1.0	1.6	31	46	−1.55	2.9
6	ZA27 + 8vol% SiC	762.3	706.7	2.3	0.8	1.35	21	43	−1.1	3.5
7	ZA27 + 16vol% SiC	759.1	703.5	3.4	1.0	1.85	39	54	−2.49	1.3
8	ZA27 + 16vol% SiC	759.1	703.5	2.1	0.8	0.9	17	43	−1.01	3.0

The samples were melted and solidified directionally upwards in an experimental setup (see Figure 1) described elsewhere [16, 17]. The procedure was as follows: at first, the liquid metal inside the mold in the furnace was allowed to reach the selected temperature. Once a uniform temperature was reached, less than 1 degree of difference recorded between the five thermocouples located at different positions in the melt, the furnace power was turned off and the melt was allowed to solidify from the bottom. Heat was extracted using a cooling system which consisted of a copper disk attached to a copper coil.

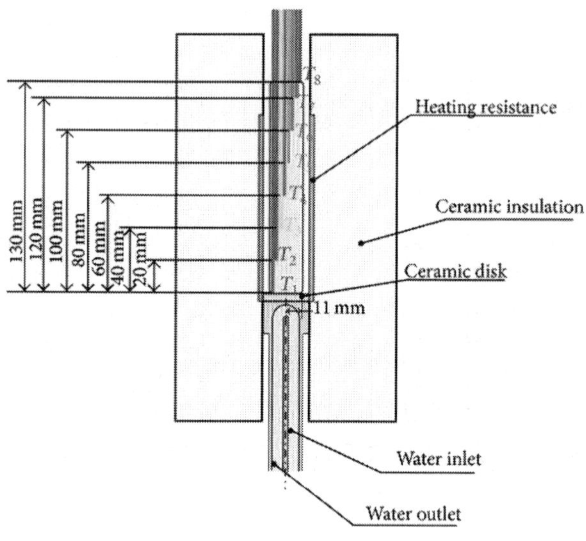

Figure 1: Details of the vertical experimental device.

The temperature measurements were performed using K-type (chromel/alumel) thermocouples (1.8 mm) which were protected with ceramic shields (i.d. 2.5 mm, E.D. 3 mm).

The thermocouples were previously calibrated using four temperature points: demineralized water at the freezing and boiling points (corrected by atmospheric pressure) and zinc (99.98 wt pct) and aluminum (99.96 wt pct) at their melting points. The accuracy of the thermocouples was determined to be about ±0.5 K. The samples were melted in ceramic molds (i.d. 2.3 mm, E.D. 2.7 mm).

The liquidus and solidus temperatures for each alloy were determined using two methods.

- The differential thermal analysis (DTA) system, NETZSCH STA 449 C with calibrated cells with pure elements was used. The DTA measurements were carried out with a charge of 200 mg of premelted samples in alumina crucibles in an Argon atmosphere. The heating and cooling velocities were 10°C/min. The liquidus (T_L) and solidus (T_S) temperatures were taken from the heating and cooling curves as usual [22, 23].
- Determining the start and the end of solidification at each thermocouple by the changes in the slopes of the cooling curve at the start and the end of solidification. This criterion was chosen to allow undercooling to occur before solidification and possible recalescence during solidification of equiaxed grains. The results between two methods (Table 1) are within 5% error and within the predicted values given by the phase diagram [24].

After solidification, the samples were cut in an axial direction, polished with emery paper up to 1000 grit and 1 μm alumina using a low speed machine, and etched with a mix containing chromic acid (50 g Cr_2O_3, 4 g Na_2SO_4 in 100 mL of water) for approximately 10 seconds at room temperature [25]. The position of the transition was located by visual observation and optical microscopy and the distance from the chill zone of the sample was measured with a ruler.

The average grain size and volume fraction were determined according to the ASTM E 112-88 and ASTM E 562-89 techniques, respectively [26, 27]. The size, volume, and number of particles in three dimensions were determined using Saltykov's modification of Johnson's method [28]. The number of particles was determined by means of a grid method, dividing one section in 64 squares of 20 × 20 mm uniformly distributed. The number of repetitions in each case ensured a representative distribution in each sample. The particle volume distribution was obtained using the standard norm ASTM E562-89. The density of average sizes of particles was determined by quantifying the repetitions of size of different particles in the grid. An average of the distribution of sizes was obtained. Particle size ranged between 1.56 μm and 20.79 μm of average diameter.

The microstructure was analyzed with optical and scanning electron microscopy (SEM). The distribution of elements in the microstructure was determined using EDXA. A Rigaku X-ray diffraction (XRD) system (Rigaku MSC, the Woodlands, TX, USA) was used for the XRD analysis of the alloys and composites.

RESULTS AND DISCUSSION

Macrostructures and Microstructures

Eight experiments were performed with different alloy and composite compositions and cooling rates. The typical macrostructures of the transition for (a) ZA27+5 vol% SiC, (b) ZA27+8 vol% SiC, and (c) ZA27+16 vol% SiC are shown in Figure 2.

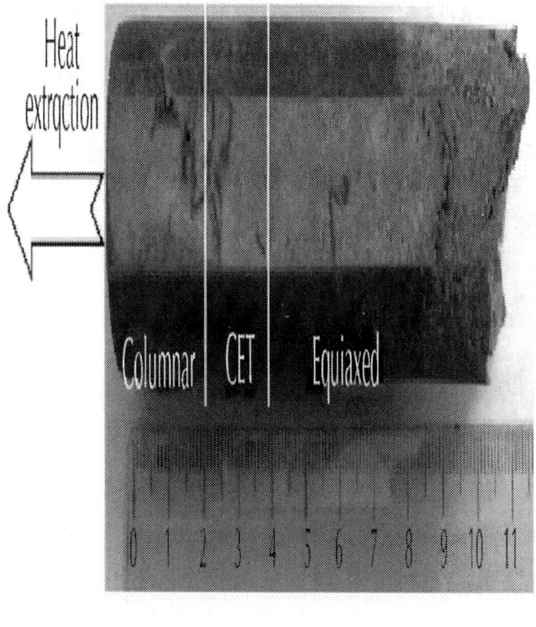

(a)

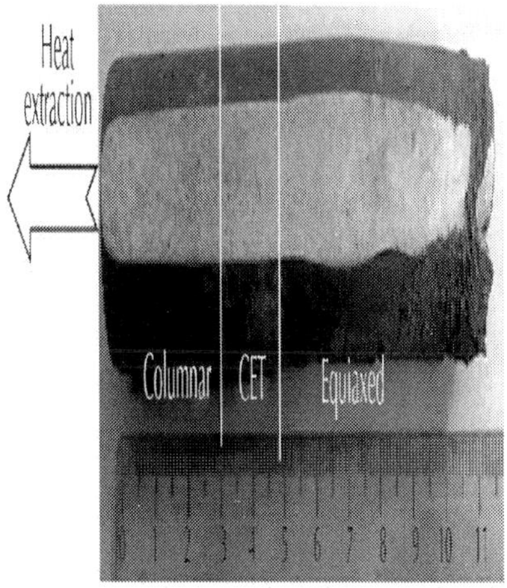

(b)

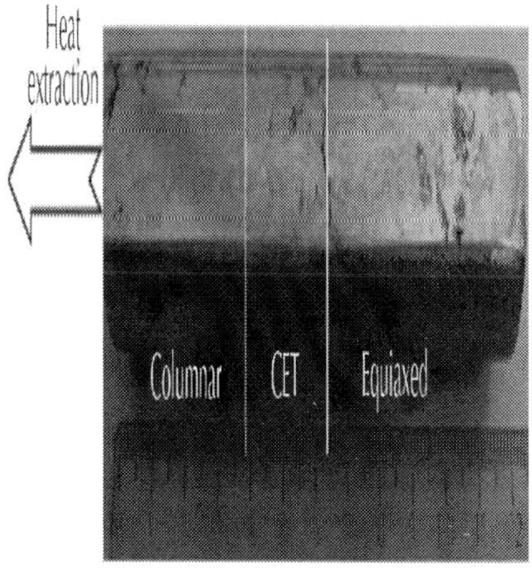

(c)

Figure 2: CET for three different composites: (a) ZA27+5 vol% SiC, (b) ZA27+8 vol% SiC, and (c) ZA27+16 vol% SiC.

As reported elsewhere for other alloys [10, 11, 16, 17], the transition is not sharp, showing a zone where some equiaxed grains coexist with columnar grains. The size of the transition zone is in the order of up to ten millimeters between the minimum position of the CET ($CET_{MIN.}$) and the maximum position of the CET ($CET_{MAX.}$). Also, no effect of the set of the thermocouples in the transition is observed; that is, the set of the thermocouples in the transition neither acted as nucleating sites nor changed the solidification structure.

The solidification microstructure of Zn-27 wt.% Al (ZA27) alloy without the presence of particles (Figure 3) presented a dendritic structure consisting of primary dendrites rich in aluminum. The $\alpha+\eta$ eutectoid is formed during the final stage of solidification, from α dendrites and β peritectic through a transformation at 548 K, following the Zn-Al phase diagram. This $\alpha+\eta$ eutectoid has a typical plate-like form of α and β sheets and a standard and finer eutectic structure. With the addition of silicon to the alloy, the phases and microstructures were the same as those in ZA27 alloys. However, silicon in the microstructure appears in two forms: in solution and as a precipitate [16]. As an example, in Figures 4(a), 4(b), and 4(c) the distributions of Si, Al and Zn in the matrix of Figure 4(c) are shown.

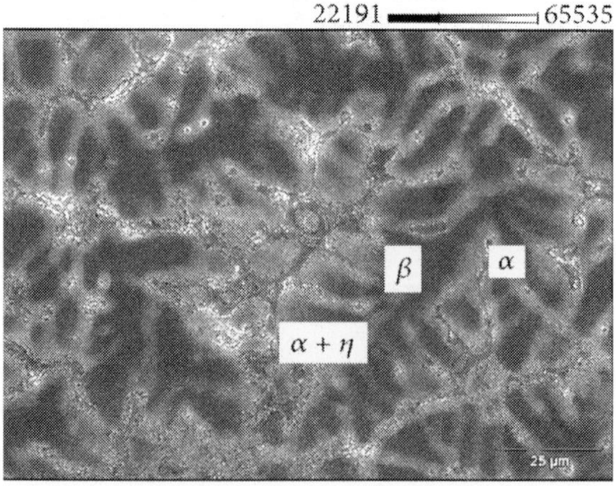

(a)

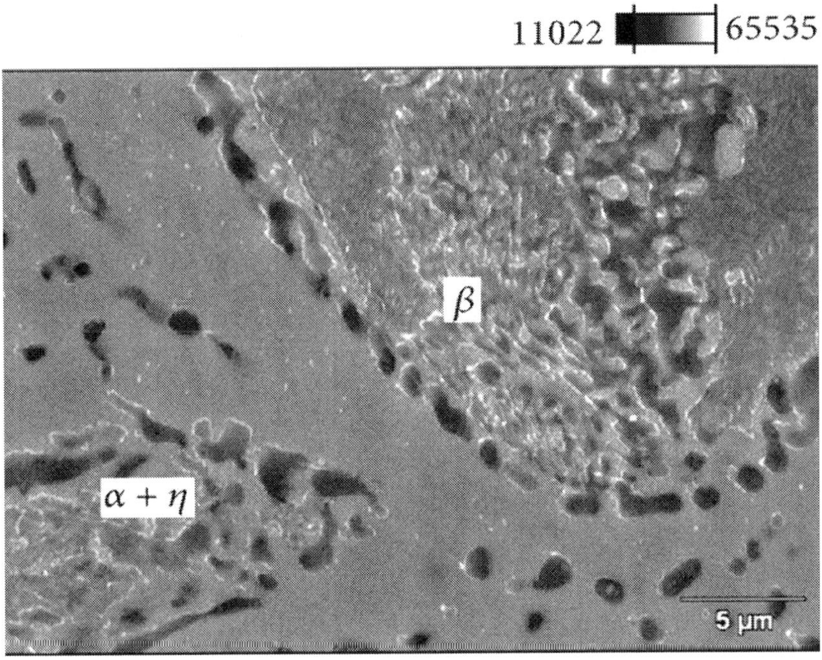

(b)

Figure 3: Microstructure of a Zn-27 wt.% Al (ZA27) alloy. The α+η eutectoid is formed during the final stage of solidification, from α dendrites and β peritectic through a transformation at 548 K, following the Zn-Al phase diagram.

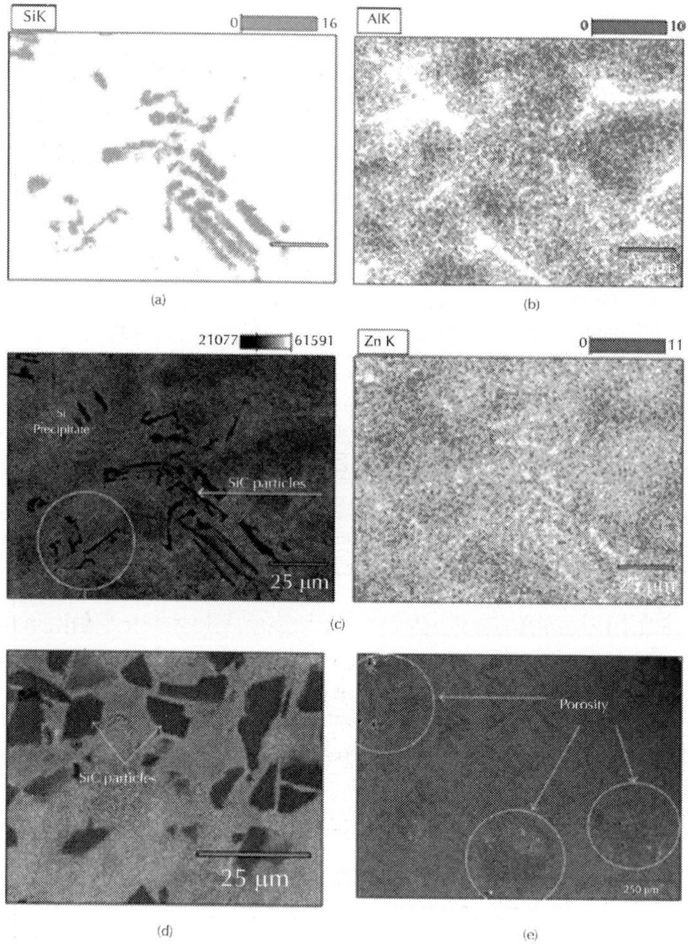

Figure 4: Microstructure of ZA27+5 vol% SiC, showing the distribution of elements, precipitates, and SiC particles.

Figure 4(c) shows the matrix with silicon precipitates and with SiC particles, respectively. The precipitates appear mostly in the dendritic cores, which indicates that the precipitates form in the melt and the grains possibly nucleate on them. Moreover, the particles appear mostly in the interdendritic regions as opposed to the silicon precipitates, as shown in Figure 4(d).

The SiC particles are clearly distinguished from the silicon precipitates because the precipitates have a light gray color, whereas the particles have a dark gray color, and because the precipitates have irregular contours, whereas the particles have sharp corners.

The samples present some degree of porosity, which increases from the bottom to the top of the casting, following the increase in the number of SiC particles (Figure 4(e)). Gas porosity can sometimes be difficult to distinguish from microshrinkage because microshrinkage cavities can contain gases as well (shrinkage defects occur when feed metal is not available to compensate for shrinkage as the metal solidifies). In general, microporosities will form if the casting is not properly risered or if a material with a wide solidification range is cast. If neither of these are the case then most likely the porosity is due to gas formation. In aluminum castings, hydrogen is the only gas that dissolves in significant quantity, which can result in hydrogen gas porosity. For castings that are a few kilograms in weight the pores are usually 0.01 to 0.5 mm in size [29].

The distribution of particles along the samples is not uniform (Figure 5). In this case, three distributions, corresponding to the columnar, CET, and equiaxed zones, are shown. A higher density of smaller particles in the columnar zone, a more monotonic decreasing distribution with a larger number with larger particles in the equiaxed zone, and finally, a more uniform distribution of particles in the equiaxed zone with a more constant number of particles for each size were observed, independently of the nature of the particle and volume fraction. This shows that, as expected, the larger particles have floated in a degree which is not significant as estimated. This could be attributed to the same solidification process itself. The chance of flotation is expected to be limited to the period of columnar solidification when the melt is completely fluid. As soon as the transition starts, there are two effects which decrease the degree of flotation: one associated with the equiaxed grains which start to solidify almost homogeneously considering the fast advance of the liquidus front as shown before; and an increase in melt viscosity (as well as in effective viscosity) due to the low temperature of the melt. Both conditions lead to a lower mobility of the particle and to a decrease in melt viscosity due to the temperature and increase in tortuosity and to the increase in the effect of viscosity due to the presence of the equiaxed grains.

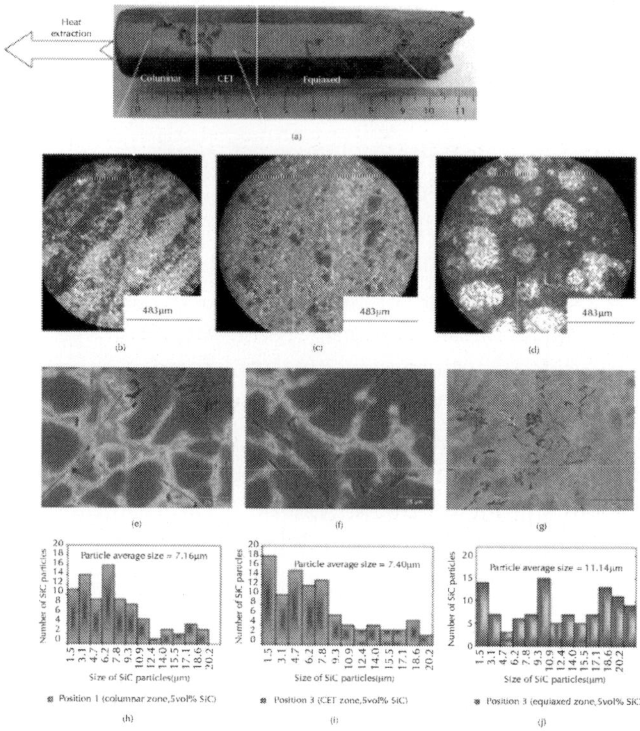

Figure 5: (a) CET for ZA27+5 vol% SiC composite. Also, micrographs in each zone: ((b) and (e)) columnar, ((c) and (f)) CET, ((d) and (g)) equiaxed. (h)–(j) Particle distribution in columnar, CET, and equiaxed zones.

The effect of the presence of foreign ceramic particles in the ZA alloys can be seen by analyzing the grain size distribution along the samples shown in Figures 6(a) and 6(b), which correspond to the grain size as a function of the position in the sample, along the direction of solidification, including the three zones (columnar, CET, and equiaxed). In the case of the columnar grains, the dimension shown is the columnar width perpendicular to the solidification direction.

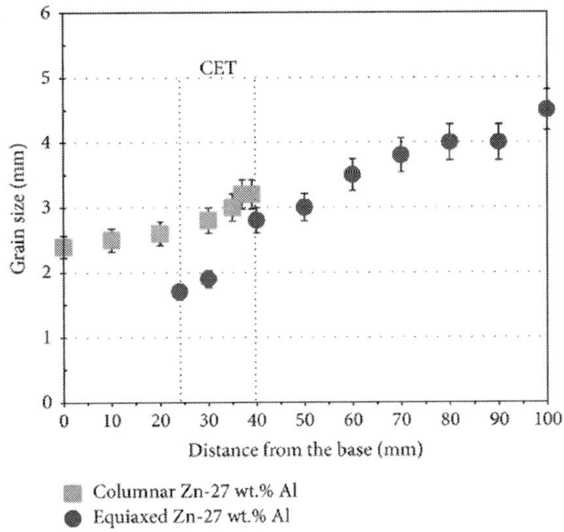

(a)

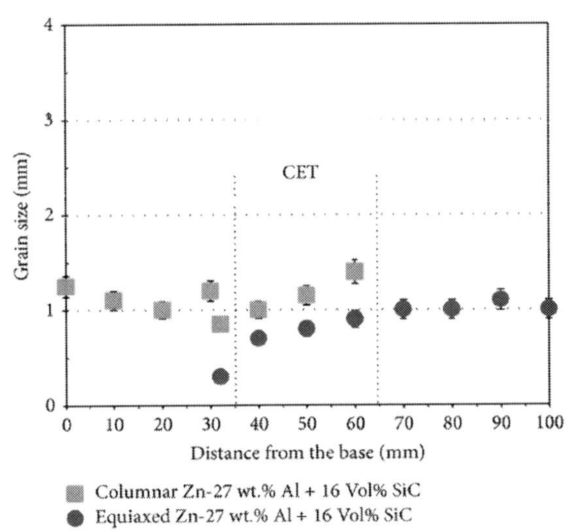

(b)

Figure 6: Grain size versus distance from the bottom of the sample. (a) Zn-27 wt.% Al.; (b) ZA27+16 vol% SiC.

The size distributions shown correspond to the Zn-27 wt.% Al alloy and the same alloy with SiC particles 16% in volume of particles. The first observation made is that the addition of particles in all cases caused a reduction in grain size in the three zones. The degree of reduction depends on the zone, the particle density, and the nature of the particle. In addition, the transition changed from columnar to equiaxed. Focusing on the columnar zone, it is observed that the SiC particles lead to a two- to three-fold decrease in columnar width, which increases with volume particle. This effect is independent of the number of particles added.

During the transition, the size of the equiaxed grains is always smaller than the columnar width, as observed in other alloys [16, 17]. Moreover, as in most cases, size increases after the transition until a maximum is reached and then normally decreases slightly again until the end of solidification.

However, the addition of particles produces a finer structure than in the columnar zone and the equiaxed grain size is never larger than the columnar width, as opposed to the case observed for the ZA27 alloy in Figure 6(a), where the grains are as large as 5 mm and the composites are rarely larger than 1 mm, independently of the particle volume fraction.

This indicates a large effect of particle addition in the resulting microstructure. This result may be caused by a number of effects related to the particular solidification conditions during nucleation, solidification, and growth, which are both affected by the particles in the light of the results. The finer structure caused by particle addition is in concordance with the results obtained by Auras and Schvezov [4] and by Lo et al. [30] and opposite to those of Karni et al. [31].

It is interesting to note that the macroscopic heat extraction rate is similar for all 8 experiments; at the moment of the transition, the value of the heat extraction, $Q_{Crit.}$, is between 1200 and 3100 W/m^2. This indicates that the effect producing a finer structure is mostly local. A finer structure may be produced by a nucleating agent and/or a larger supercooling.

The first situation is unlikely because the ceramic particles do not promote nucleation due to the surface tension which produces an unfavorable thermodynamic condition increasing the free energy. This is also supported by the observation that most of the particles appear in the interdendritic regions and not inside the dendrite trunks, which would be the situation if the particles act as nucleation sites for the equiaxed grains.

Temperature Measurements

The typical time-dependent temperatures measured by the thermocouples in each position of the samples with a CET are shown in Figure 7 for ZA27 +16 vol% SiC.

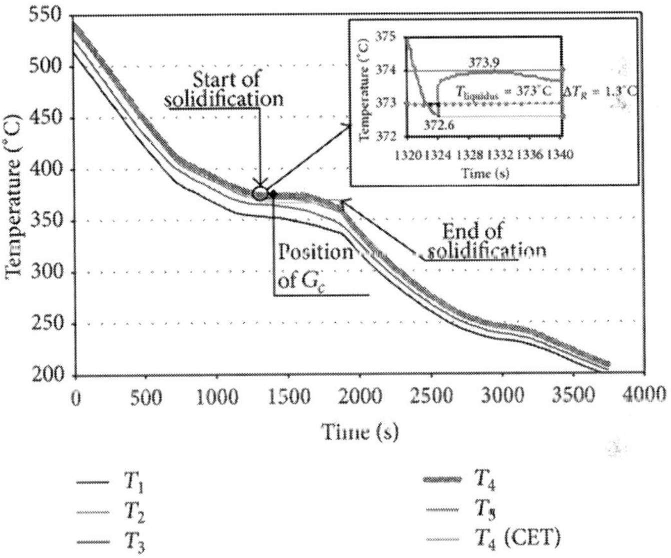

Figure 7: Cooling curve for ZA27 +16 vol% SiC composite showing positions of critical temperature gradient for one thermocouple location (CET position), recalescence, and start and end of solidification.

In the figure, the thermocouple T_1 is at the lowest position, and is the first to reach the solidification front, whereas T_8 is at the highest position. In all the curves, it is possible to identify a period corresponding to the cooling of the melt, a second period of solidification, and a final period of cooling of the solid to ambient temperature.

Solidification Parameters

The data shown in Figure 7 allow extracting the following information: melt superheat, cooling rate of the melt and the solid, position of the solidification fronts for the solidus and the liquidus temperatures, local solidification time, velocity of both solidification fronts, length of the mushy zone, temperature gradients and recalescence.

In some cases, the quantification of these parameters is straightforward. This is the case of melt superheat, which is the difference between the highest temperature above the liquidus reached by the melt before the furnace is turned off and the liquidus temperature. It is also the case of the cooling rate, which is calculated as the slope of the temperature versus time curve in both periods, the cooling of the melt and the solid. The method used to calculate the parameters from the temperature versus time curves has been reported before [10–13, 16, 17].

The change in slope of the temperature versus time curve measured for the liquidus temperature, T_L, or the start of solidification was typically around 2.1 K/s, whereas that for the solidus temperature, T_S, or the end of solidification was around 0.9 K/s. Both T_L and T_S are given in Table 1.

Cooling Rate

The cooling rate in the liquid alloy was determined from the temperature versus time curves at each thermocouple position, which was calculated as the slope of the temperature curve in both periods that refer to (a) the cooling rate of the melt (liquid), T_L, and (b) the cooling rate of the solid, T_S.

The temperature versus time plot for one experiment is presented in Figure 7. The cooling rate in the melt, , and the cooling rate of the solid, , both calculated from this type of curves are listed in Table 1 for all the experiments; velocities of 2.0 K/s to 3.5 K/s were produced. Table 1 also lists the location of the transition from the bottom of the sample which is in the range of $CET_{MIN.}$ to $CET_{MAX.}$ (cm). Comparing the cooling velocities with the distances which correspond to the length of the columnar zone, it is observed that the length of the columnar grains (CET_{MAX}) increases as the velocity increases.

The temperature versus time curves also show that the temperature evolution depends on the structure being formed. During columnar solidification, the temperature decreases steadily and monotonically, whereas in the equiaxed region, during the CET, there is a recalescence which increases the temperature from a minimum; the level of recalescence for each experiment is listed in Table 1 as R (°C) and it is shown in the inset of Figure 7 for experiment number 7.

Solidification Velocity

The solidification velocity was determined from the time recorded from the start to the end of solidification at each thermocouple position (the separation between thermocouples was 20 mm). The position of the interphases for one experiment with composites is shown in Figures 8(a) and 8(b). A uniform motion of the liquidus and solidus interphases is observed as in the case of Zn-Al alloys reported before [13]. The horizontal bar in each position indicates the time it takes the temperature to go from the liquidus to the solidus temperature, that is, the local solidification time versus liquidus and solidus interphases. The velocities can easily be calculated from these types of figures or from Figure 9 for the same experiments.

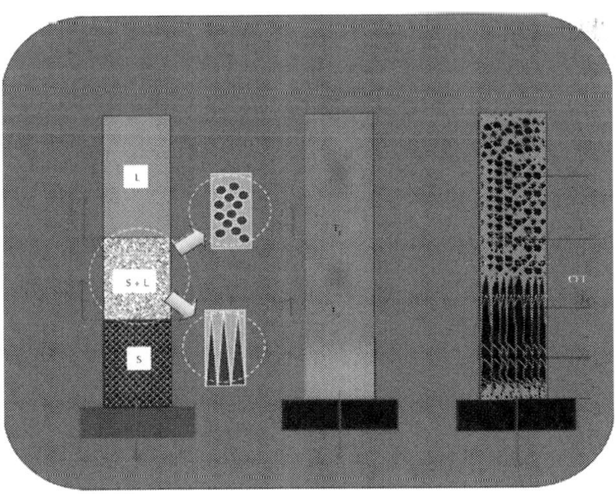

(a)

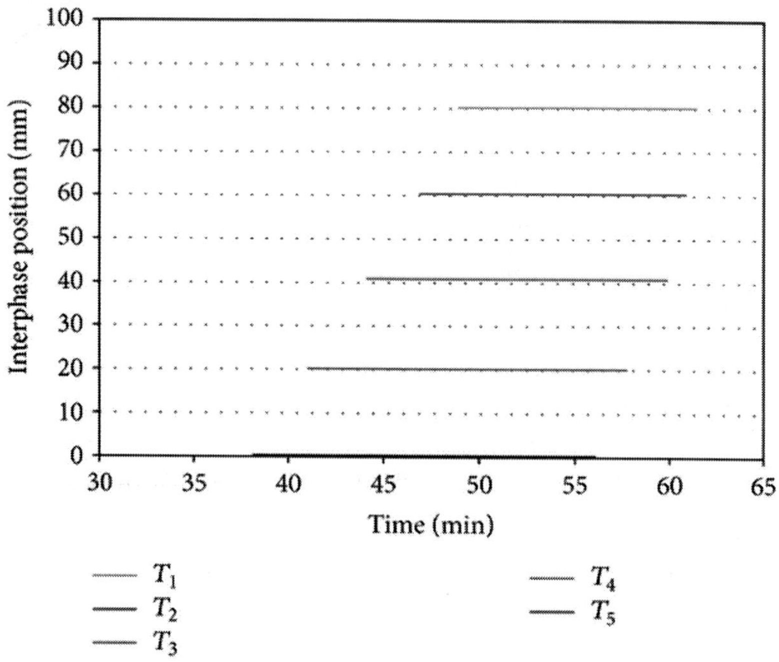

(b)

Figure 8: Interphase position versus time during solidification of composites. (a) Schematic and (b) ZA27+16 vol% SiC.

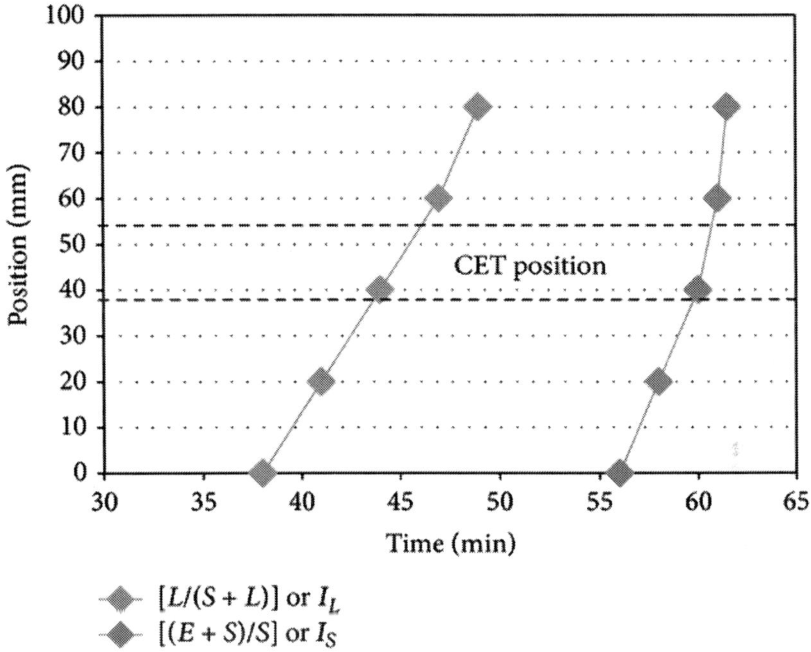

Figure 9: Liquid (L) and solid (S) interphase positions. ZA27+16 vol% SiC.

In both cases, an acceleration of the interphase movement, which becomes particularly relevant at the transition from columnar to equiaxed solidification, can be observed. It is also observed that the liquidus interphase accelerates faster than the solidus one. The numerical values of both velocities at the transition are listed in Table 2.

Table 2: Liquid (V_L) and solid (V_S) interphase velocity and local solidification velocity (V_{SL})

Experiment number	Composite	V_L (mm/s)			V_S (mm/s)			V_{SL} (mm/s)		
		V_L^1	V_L^2	V_L^3	V_S^1	V_S^2	V_S^3	V_{SL}^1	V_{SL}^2	V_{SL}^3
3	ZA27 + 5vol% SiC	0.02	0.15	0.21	0.02	0.03	0.06	−1.00	0.14	0.13
4	ZA27 + 5vol% SiC	0.14	0.19	0.40	0.07	0.09	−0.14	0.17	0.16	0.10
5	ZA27 + 8vol% SiC	0.09	0.09	0.44	0.12	0.04	0.14	0.14	0.09	0.21
6	ZA27 + 8vol% SiC	0.04	0.07	0.23	0.04	0.07	0.14	−1.49	−1.13	−1.07
7	ZA27 + 16vol% SiC	0.03	0.24	0.27	0.02	0.10	0.14	0.07	0.17	0.30
8	ZA27 + 16vol% SiC	0.07	0.33	0.03	0.06	0.08	0.06	0.41	0.08	−0.08

The position of the interphase as a function of the difference in time for each thermocouple is used to define a parameter which is called the local solidification velocity (V_{SL}) using the following equation [18]:

$$V_{SL} = \frac{V_L * V_S}{V_L - V_S}. \tag{1}$$

It is possible that this local solidification velocity acquires subsequent values:

- $V_L = V_S \Rightarrow V_{SL} \rightarrow \infty,$
- $V_{SL} > 0 \Rightarrow V_L > V_S,$
- $V_{SL} < 0 \Rightarrow V_L < V_S,$
- $V_{SL} = 0 \Rightarrow V_L = V_S = 0.$

In Table 2, V_{SL} is negative between the first two thermocouples at the bottom, indicating that the solid interphase moves faster than liquid interphase, reducing the size of the mushy zone. After this, the liquid interphase accelerates faster than the solidus one, giving positive values for V_{SL} and increasing the size of the mushy zone.

Figure 10 shows the different velocities, V_L, V_S, and V_{SL}, obtained for one of the experiments. In this case, both the velocities of the solid and liquid interphase accelerate during and after the transition, resulting in a V_{SL} which also increases drastically, leading to a larger mushy zone. The highlighted boxes in Table 2correspond to the liquidus velocities obtained at the critical point of the CET(V_{LC}). The observation here is that these values are nearly the largest obtained in each experiment, indicating the acceleration of the liquidus interphase at the transition.

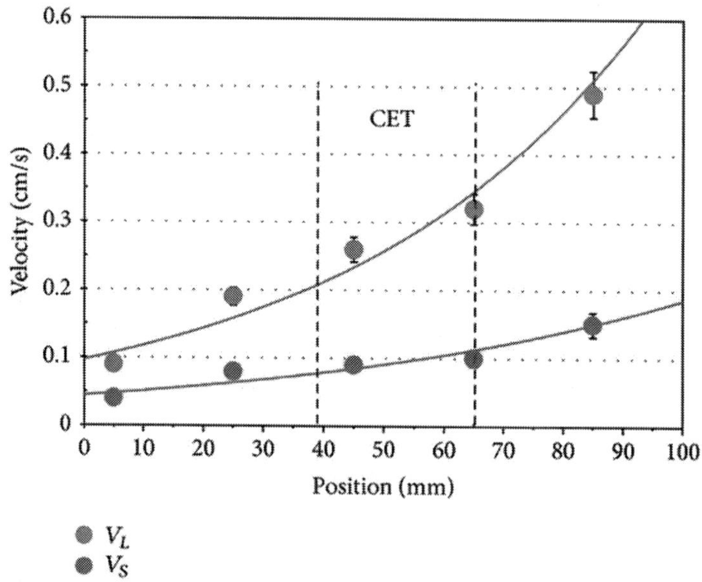

Figure 10: V_L,V_S and V_{SL} versus time. ZA27+16 vol% SiC.

Temperature Gradients

The temperature gradients, G, were calculated for each pair of neighboring thermocouples as the temperature difference between the thermocouple readings divided by the separation distance between thermocouples.

The values of gradients are plotted in Figure 11 for one experiment and for each type of composite. In all the figures it is observed that from the beginning of solidification, the gradients decrease with the time. The minimum value always corresponds to the position of the CET. The gradients shown for the experiment in Figure 11(a) reach negative values at this point (−2.49°C/cm). This negative value indicates a reversal in the temperature profiles ahead of the interface, which could be associated with the recalescence due to massive nucleation of equiaxed grains, as previously reported and discussed for different alloys [10–13, 16, 17]. The critical temperature gradients for all experiments are presented in Table 1.

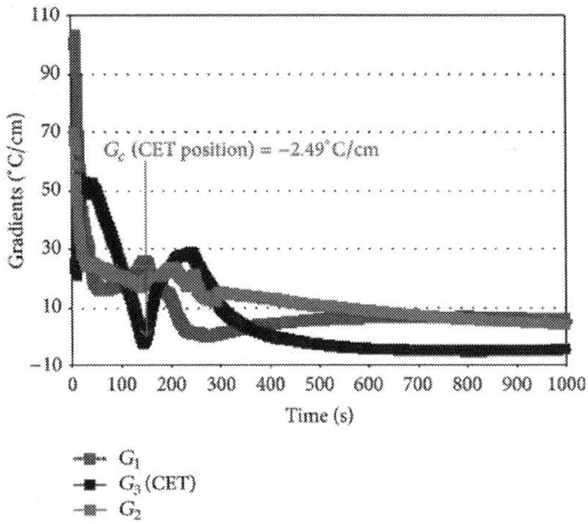

Figure 11: Temperature gradients versus time for ZA27+16 vol% SiC.

Semiempirical Modeling

The ceramic particles are more likely to act as retardants of the nucleation processes. This view is supported by the following facts associated with the dynamics during and after the transition. The two key parameters are the critical velocity of the liquidus front during the transition, V_{LC}, and the temperature gradient in the region, G_C.

First, the V_{LC} is associated with the kinetics of nucleation of the equiaxed grains; V_{LC} in all cases accelerates very rapidly during and after the transition. G_C indicates the degree of massivity and severity of the conditions present during the equiaxed nucleation, that is, the degree of supercooling necessary for nucleation and the recalescence product of the solidification process.

By comparing these two parameters for the ZA27 alloys with those for the composites, it is seen that the V_{LC} for the former is between 1.5 and 1.9 mm/s and the average gradient measured is slightly negative but small. On the other hand, the reinforced alloys produce a V_{LC} that is similar and an average temperature gradient that is always negative. This indicates that the conditions required for nucleation in the presence of ceramic particles are more severe and that the kinetics of solidification once the transition starts is not affected because is similar in all cases.

Another parameter associated with the nucleation process is the recalescence; by comparing the recalescence for ZA27 and that for its composites, it is observed that there is no significant increase in the recalescence for the composites. The recalescence is associated with the mass of alloy that nucleates and not with the number of nuclei or the amount of heat that may be absorbed by the surrounding melt, which in turn is related to the amount of supercooling. Time-temperature measurements showed that recalescence occurs before the critical temperature gradient was reached (see Figure 7).

The overall solidification process is controlled and limited by the heat extraction from the solidification region given by Q_{Crit}., which is similar for all cases.

On the other hand, the experimental observations and measurements show that the transition in the composites occurs when the temperature gradient reaches values below 0.1 K/mm and the liquidus front velocity increases to values around 0.2 to 0.4 mm/s. Following these two conditions, the position of the transition may be located at any stage of solidification. Both the gradient and liquidus front velocities can be calculated using numerical techniques [8]. The calculation of the thermal field in Stefan's problem was done using an explicit finite difference method to solve the unidirectional heat flow equation. The latent heat of the solidifying material in the mushy zone was calculated using the enthalpy method [32, 33]. The specific heats and the thermal conductivities of solid and liquid were assumed as temperature and concentration dependent. The resulting nonlinear equations were linearized employing a modified temperature approach. The heat transfer coefficients at the bottom and top of the sample were determined as a function of temperature by inverse modeling.

The positions of the CET in MMCs predicted for the eight experiments are shown in Table 3. As it can be observed, the agreement between the model and the experimental results is reasonably good. The largest discrepancy is lower than 10%.

Table 3: Predicted and measured CET positions for different composites

Experiment number	Alloy and composite	$CET_{predicted}$ (mm)	$CET_{measured}$ (mm)	Relative error (%)
1	Zn-27 wt.% Al (ZA27)	91	86	5.49
2	Zn-27 wt.% Al (ZA27)	47	45	4.25
3	ZA27 + 5vol% SiC	52	47	9.61
4	ZA27 + 5vol% SiC	69	65	5.79
5	ZA27 + 8vol% SiC	50	47	6.02
6	ZA27 + 8vol% SiC	46	42	8.69
7	ZA27 + 16vol% SiC	57	53	7.01
8	ZA27 + 16vol% SiC	45	42	6.66

CONCLUSIONS

The main results obtained in the present investigation may be summarized as follows.

- Metal-matrix composites were directionally solidified and the temperatures during the whole process were measured in the liquid and solid phases.
- For the three types of composites studied, the columnar-to-equiaxed transition (CET) was produced and the values of the temperature gradients calculated reached minimum values during the transition and were even negative in most cases.
- ZA27+16 vol% SiC showed a second minimum of the gradient after the transition, indicating the importance and complexity of the solidification pattern of composites with technological interest.

- In the three types of composites studied an increase in the cooling velocity in the liquid was found to increase the columnar zone length.
- At the CET, the interphase velocities are high (between 0.07 and 0.44 mm/s) depending on the experiment.
- Recalescence was detected and measured during the equiaxed transition, being of the order of 3.5°C to 1.1°C.
- The CET structure is not abrupt but occurs in a zone of 1 cm or larger. The same results were obtained for the binary zinc-aluminum alloys.
- The presence of ceramic particles in ZA27 alloys affects the thermodynamic local conditions and the kinetics of nucleation, producing a finer microstructure.
- The position of the CET in MMCs is predicted within a 10% error.

ACKNOWLEDGMENTS

This work was partially supported by the Argentinean Research Council (CONICET) and the National Agency for Science and Technology (PICT-2011-1378).

REFERENCES

1. W. D. Callister Jr., Materials Science and Engineering: An Introduction, John Wiley & Sons, New York, NY, USA, 7th edition, 2006.
2. K. K. Chawla, Composite Materials: Science & Engineering, Springer, New York, NY, USA, 2nd edition, 1999.
3. K. K. Chawla and M. Metzger, Advances in Research on Strength and Fracture of Materials, vol. 3, Pergamon Press, New York, NY, USA, 1978.
4. R. Auras and C. Schvezov, "Wear behavior, microstructure, and dimensional stability of as-cast zinc-aluminum/SIC (metal matrix composites) alloys," Metallurgical and Materials Transactions A, vol. 35, no. 5, pp. 1579–1590, 2004.

5. J. D. Hunt, "Steady state columnar and equiaxed growth of dendrites and eutectic," Materials Science and Engineering, vol. 65, no. 1, pp. 75–83, 1984.
6. R. B. Mahapatra and F. Weinberg, "The columnar to equiaxed transition in tin-lead alloys,"Metallurgical Transactions B, vol. 18, no. 2, pp. 425–432, 1987. · ·
7. I. Ziv and F. Weinberg, "The columnar-to-equiaxed transition in Al 3 Pct Cu," Metallurgical Transactions B, vol. 20, no. 5, pp. 731–734, 1989. · ·
8. C.-A. Gandin, "Experimental study of the transition from constrained to unconstrained growth during directional solidification," ISIJ International, vol. 40, no. 10, pp. 971–979, 2000.
9. C.-A. Gandin, "From constrained to unconstrained growth during directional solidification," Acta Materialia, vol. 48, no. 10, pp. 2483–2501, 2000. · ·
10. A. E. Ares and C. E. Schvezov, "Solidification parameters during the columnar-to-equiaxed transition in lead-tin alloys," Metallurgical and Materials Transactions A, vol. 31, no. 6, pp. 1611–1625, 2000.
11. A. E. Ares, S. F. Gueijman, and C. E. Schvezov, "Semi-empirical modeling for columnar and equiaxed growth of alloys," Journal of Crystal Growth, vol. 241, no. 1-2, pp. 235–240, 2002. · ·
12. A. E. Ares, S. F. Gueijman, R. Caram, and C. E. Schvezov, "Analysis of solidification parameters during solidification of lead and aluminum base alloys," Journal of Crystal Growth, vol. 275, no. 1-2, pp. e319–e327, 2005. · ·
13. G. Reinhart, N. Mangelinck-Noël, H. Nguyen-Thi et al., "Investigation of columnar-equiaxed transition and equiaxed growth of aluminium based alloys by X-ray radiography," Materials Science and Engineering A, vol. 413-414, pp. 384–388, 2005. · ·
14. H. Yasuda, I. Ohnaka, A. Sugiyama et al., "Time-resolved imaging of the microstructure evolution during dendritic solidification by using synchrotron radiation," in Modeling of Casting, Welding and Advanced Solidification Processes—XI, C.-A. Gandin and M. Bellet, Eds., pp. 375–382, TMS, Warrendale, Pa, USA, 2006.

15. J. A. Spittle, "Columnar to equiaxed grain transition in as solidified alloys," International Materials Reviews, vol. 51, no. 4, pp. 247–269, 2006. · ·
16. A. E. Ares and C. E. Schvezov, "Influence of solidification thermal parameters on the columnar-to-equiaxed transition of aluminum-zinc and zinc-aluminum alloys," Metallurgical and Materials Transactions A, vol. 38, no. 7, pp. 1485–1499, 2007. · ·
17. S. F. Gueijman, C. E. Schvezov, and A. E. Ares, "Vertical and horizontal directional solidification of Zn-Al and Zn-Ag diluted alloys," Materials Transactions, vol. 51, no. 10, pp. 1861–1870, 2010. · ·
18. A. E. Ares and C. E. Schvezov, "Metal matrix composites directionally solidified," in TMS 2013 Annual Meeting Supplemental Proceedings, pp. 1061–1068, TMS, John Wiley & Sons, Warrendale, Pa, USA, 2013.
19. K. U. Kainer, "Basics of metal matrix composites," in Metal Matrix Composites: Custom-Made Materials for Automotive and Aerospace Engineering, K. U. Kainer, Ed., Wiley-VCH GmbH & Co. KGaA, Weinheim, Germany, 2006.
20. A. Mortensen and M. C. Flemings, "Solidification of binary hypoeutectic alloy matrix composite castings," Metallurgical and Materials Transactions A, vol. 27, no. 3, pp. 595–609, 1996.
21. N. F. Dean, A. Mortensen, and M. C. Flemings, "Liquid-state processing," in Fundamentals of Metal-Matrix Composites, P. K. Rohatgi, Ed., pp. 3–22, TMS, Warrendale, Pa, USA, 1993.
22. R. F. Speyer, Thermal Analysis of Materials, Marcel Dekker, New York, NY, USA, 1994.
23. Y. T. Zhu, J. H. Devletian, and A. Manthiram, "Application of differential thermal analysis to solid-solid transitions in phase diagram determination," Journal of Phase Equilibria, vol. 15, no. 1, pp. 37–41, 1994. · ·
24. W. J. Moffatt, The Handbook of Binary Phase Diagrams, General Electric Company Corporate Research and Development Technology Marketing Operation, New York, NY, USA, 1984.
25. G. F. Vander Voort, Metallography Principles and Practice, ASM International, New York, NY, USA, 2000.

26. E112-96el: ASTM Standards, vol. 03.01, ASTM, Philadelphia, Pa, USA, 1988.
27. E562-02: ASTM Standards, vol. 03.01, ASTM, Philadelphia, Pa, USA, 1989.
28. E. E. Underwood, "Particle size distribution," in Quantitative Microscopy, R. T. de Hoff and F. N. Rhines, Eds., pp. 76–101, 149–199, McGraw-Hill, New York, NY, USA, 1968.
29. B. Prillhofer, H. Böttcher, and H. Antrekowitsch, "Development and practical performance characteristics of a new impeller for metal treatment in casting/holding furnaces," in Light Metals 2009, p. 234, The Minerals, Metals & Materials Society, 2009.
30. S. H. J. Lo, S. Dionne, M. Sahoo, and H. M. Hawthorne, "Mechanical and tribological properties of zinc-aluminium metal-matrix composites," Journal of Materials Science, vol. 27, no. 21, pp. 5681–5691, 1992.
31. N. Karni, G. B. Barkay, and M. Bamberger, "Structure and properties of metal-matrix composite,"Journal of Materials Science Letters, vol. 13, no. 7, pp. 541–544, 1994.
32. S. F. Gueijman, A. E. Ares, and C. E. Schvezov, "Enthalpy variations and latent heat evolution during solidification of lead tin alloys," in Light Metals 2000, pp. 615–621, TMS, Warrendale, Pa, USA, 2000.
33. A. E. Ares, S. F. Gueijman, and C. E. Schvezov, "Latent heat evolution during solidification of aluminum based alloys," in Light Metals 2001, pp. 1077–1084, TMS, Warrendale, Pa, USA, 2001.

Chapter 4

Effect of Process Parameters on Gas Nitriding of Grey Cast Iron

Nana Wang and Jinxiang Liu

School of Mechanical Engineering, Beijing Institute of Technology, Beijing 100081, China

ABSTRACT

Gas nitriding process parameters have significant effects on the nitriding layer of material. In the present work, a series of gas nitriding experiments on pearlitic grey cast iron specimens were carried out at different temperatures. To study the influence of nitriding process parameters on the nitriding layer, the numerical simulations of nitriding processing are performed, which take into account the threshold value of nitriding potential. The results show that the numerical simulations, incorporating the nitriding potential's threshold value, can accurately predict the case depth and nitrogen concentration profile. The nitriding layer increases with increasing nitriding time and temperature, whereas the nitrogen concentration on the surface decreases with increasing temperature. Besides, the results also reveal that the nitriding potential nearly has no effect on the case depth, whereas it has great influence on nitrogen concentration on the surface and chemical composition of the compound layer.

INTRODUCTION

Gas nitriding is a typical thermochemical surface treatment within the eutectoid temperature, in which the nitrogen is transferred from the ammonia gas into the workpiece surface, generating the gamma phase (γ'-Fe_4N) and epsilon phase (ε-Fe_3N) [1, 2]. After nitriding, the so-called compound layer and diffusion zone can be formed, which are capable of improving the wear and corrosion resistance of materials and enhancing the fatigue endurance [3, 4], respectively. The nitriding process parameters, such as nitriding time, nitriding temperature, and nitriding potential, have great influence on the forming of nitriding layer. The nitriding process parameters are mainly determined by experience in practical engineering applications. Their effects on nitriding layer are not very clear from a mathematical view. Therefore, considering the sensitivity of nitriding layer to the process parameters, there is an urgent demand to investigate the influence of the process parameters mathematically.

In recent years, a significant amount of experimental work has been conducted, aiming to investigate the nitriding process parameters' effect on nitriding process. Nevertheless, due to high time-consuming and economic costs, the experimental method is not very effective and feasible in practical applications, especially when the parameters vary in a wide range. Under the circumstances, numerical analysis can be a good choice to study the processing parameters' influence. Arif et al. [5] investigated the effect of nitriding time, nitriding temperature, and nitriding potential on the nitrogen concentration and hardness depth profile by using finite element code. In their research, the influence of the compound layer on the nitrogen concentration distribution was not considered, so the calculated case depth was larger than the experimental data. To avoid the aforementioned shortcomings, Yang et al. [6, 7] proposed a compound layer growth model for an alloy steel. However, the difference between the critical value and threshold value of nitriding potential was ignored in Yang's research. Besides, Hu et al. [8, 9] simulated the nitriding process under multistage nitriding potential controlling. In their study, a nitriding layer without the compound layer was obtained by controlling nitriding process parameters. Furthermore, most of the researches published in the literature were focused on the nitriding process of pure iron or steels. For grey cast iron, a widely used material, related studies are lacking.

In the present work, a series of gas nitriding experiments on grey cast iron specimens were conducted at 550, 570, and 590°C. Afterwards, the numerical simulations were performed, which combined the influence of nitriding potential's threshold value on nitriding layer and Fick's second law. The calculated thickness of diffusion zone was compared with the tested data to verify the numerical simulations. Finally, the effect of nitriding process parameters, including nitriding temperature, nitriding potential, and nitriding time, was further investigated.

EXPERIMENTAL WORK

Material

The material tested in the study is pearlitic grey cast iron. The chemical composition of the material is shown in Table 1. Figure 1 depicts the microstructure of the grey cast iron. The graphite within the material is in a sheet form. Due to the large amount of graphite, the grey cast iron has good heat transfer ability, self-lubricating performance, and good casting property. It is a very suitable material for components operating under high temperature and friction condition.

Table 1: Chemical composition (wt. %) of the studied grey cast iron

C	S	Si	P	Mn	Cr	B	Cu	Ni	Ti	V	Mo	Nb	Fe
2.98	0.032	1.646	0.061	0.469	0.051	0.003	0.348	1.278	0.045	0.042	0.874	0.066	Bal.

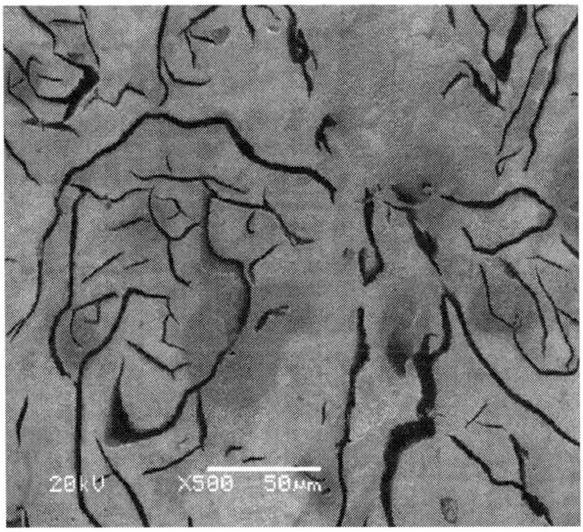

Figure 1: SEM analysis of the grey cast iron used in an engine cylinder liner.

Nitriding Process, Sample Preparation, and Equipment

Gas nitriding experiments were carried out on grey cast iron specimens. The specimens were divided into three groups, with two specimens in each group. The tested temperatures were set as 550, 570, and 590°C for each group, with the same total time of 24 hours and nitriding potential of 1.5 $atm^{-1/2}$.

After nitriding, the specimens were prepared by a standard grinding and polishing procedure, and microstructure observation was performed using JSM-5610LV SEM. Specimens were etched by 5% Nital in order to observe the nitriding layer by OLYMPUS PMG3. Hardness of the samples was also tested by applying a load of 1 N with holding time of 10 s. Besides, the hardness tests were conducted in 6 points of one sample and the average value was taken to represent the final hardness.

In order to obtain the composition of the nitriding layer, XRD analysis of the surface layer was performed using D8 ADVANCE X-ray diffractometer with CuK, at 40 kV and 40 mA. The scanned area was in the range of 15–90°.

NUMERICAL MODEL OF GAS NITRIDING PROCESS

Theoretical Background

In gas nitriding, the atomic nitrogen is dissociated from ammonia (NH_3) gas and then diffuses into the workpiece surface of component. The dissociation obeys the following rule:

$$NH_3 = \frac{3}{2}H_2 + [N], \tag{1}$$

where [N] represents N atom dissolved in the workpiece surface.

The above equation presumes that equilibrium has been established. The activated atomic nitrogen is dissolved in the iron crystal lattices, which not only promotes the nitriding of α-Fe, but also accelerates the formation and growth of compound layer. The solution of atomic nitrogen in pure iron is relatively fast, which does not affect the nitriding rate [10]. When building numerical model, for convenience, it is assumed that the graphite has no influence on the nitriding rate, although graphite is widely distributed in the iron matrix [11]. The resulting effect of this assumption will be discussed in detail in Section 5.1.

The transformation of atomic nitrogen within grey cast iron is determined by the diffusion velocity, which is described by Fick's second law:

$$\frac{\partial c}{\partial t} = D\frac{\partial^2 c}{\partial x^2}, \tag{2}$$

where c represents the nitrogen concentration within the material and D is the diffusion coefficient of atomic nitrogen.

Numerical Model and Boundary Conditions

Nitriding potential is the measurement of nitriding ability of NH_3 gas. The phases and nitrogen concentration on the nitriding surface are closely related to the nitriding potential after local equilibrium is established. At the earliest, the well-known Lehrer diagram of pure iron, which was constructed by Lehrer through investigating the equilibrium conditions of NH_3-H_2 and Fe-N [12], was used to determine the relationship of the nitrided phases, nitrogen concentration, and nitriding potential.

The critical value of nitriding potential is defined as the turning point when the nucleation of γ′ phase occurs at certain temperature. This value can be directly obtained from the Lehrer diagram. However, in engineering applications, the nitriding potential for compound layer nucleation is not completely equal to the critical value, since a minimal value, the so-called threshold value, has to be exceeded for certain nitriding time. The theoretical equation [13] for threshold value of nitriding potential is

$$K_{\mathrm{nt}} = \frac{K_{\mathrm{nc}}}{1 - \exp\left(\beta^2 t/D\right) \operatorname{erfc}\left(\beta\sqrt{t}/\sqrt{D}\right)}, \tag{3}$$

where K_{nt} is the threshold value of nitriding potential for nitriding time t, K_{nc} is the critical value of nitriding potential, β is the mass transfer coefficient in gas-solid reaction, and erfc represents the complementary error function.

At a certain nitriding temperature, the time t_c, representing the nucleation starting time of compound layer, can be obtained by substituting the critical value of nitriding potential into (3). For accuracy, a two-stage process is defined in the current study. In the first stage, that is, $t \leq t_c$, the compound layer is not formed. In the second stage, that is $t > t_c$, the compound layer starts to nucleate and grow. The initial condition for stage I can be described as

$$t = 0, \qquad c = c_0, \tag{4}$$

where c_0 is the initial nitrogen concentration within grey cast iron.

The boundary condition for stage I is

$$x = \infty, \qquad c = 0, \qquad x = 0,$$

$$-D_d \frac{\partial c}{\partial x} = \beta\left(c_g - c_s\right), \tag{5}$$

where D_d represents the diffusion coefficient of atomic nitrogen in the diffusion zone, x is the depth from surface, c_g is the nitrogen concentration in the gas phase, and c_s is the nitrogen concentration on the surface of grey cast iron.

At the end of stage I, the compound layer begins to be formed and the nitrogen concentration in the interface between compound layer and diffusion zone is in equilibrium. In the diffusing process, the nitrogen concentration in the interface is set as unchanged. Therefore, the obtained nitrogen concentration on the surface of cast iron can be considered as the initial boundary condition for stage II. Since the thickness of compound layer is much less than that of the diffusion zone [7], the influence of the interface moving during the nitriding process is ignored in the study. Based on the hypothesis, the boundary condition for stage II is of the following form:

$$x = \infty, \qquad c = 0, \qquad x = 0,$$

$$-D_d \frac{\partial c}{\partial x} = \beta\left(c_g - c_s\right), \tag{6}$$

where c_d is the nitrogen concentration on the workpiece surface after stage I.

The parameters needed to be calibrated are the diffusion coefficient D_d and mass transfer coefficient β. For parameter D_d, it is mainly influenced by the alloying element, which is capable of reacting with the atomic nitrogen. For grey cast iron, the governing alloying element is Cr, whereas it has little effect on the diffusion coefficient when its content is lower than 5% [14, 15]. The Cr element in the analyzed grey cast iron is 0.051%, so the influence can be ignored. The diffusion coefficient is determined by the following equation [16]:

$$D_d = 0.66 \exp\left(\frac{-77900}{RT}\right), \tag{7}$$

where R is the gas constant with a value of 8.314 J/mol·K and T is the nitriding temperature.

For parameter β, it is determined by the following equation [8]:

$$\beta = 0.0621 \exp\left(\frac{-6267}{T}\right). \tag{8}$$

In practice, since the compound layer is a multiphase composition, the diffusion coefficient of atomic nitrogen in this area is hard to be determined, and the thickness of compound layer can only be determined by experimental data. Therefore, the study is aimed at predicting the thickness of diffusion zone and analyzing the influence of nitriding process parameters.

EXPERIMENTAL ASSESSMENT

Optical Microscope

The cross-sectional observation through optical microscopes was made. Figure 2 shows the case depths after nitriding at 550, 570, and 590°C. As shown in the figure, the white layer is formed on the surface which is known as the compound layer. Beneath the compound layer, there is a diffusion zone, which is not so clear to be observed. It can be found that the compound layer increases with the increase in nitriding temperature, and the average thickness values at 550, 570, and 590°C are 4.50, 9.05, and 12.9 μm, respectively. The relationship between the compound layer thickness (h) and nitriding temperature is plotted in Figure 3, which exhibits a linear characteristic; that is, $h = -110.9 + 0.2T$. .

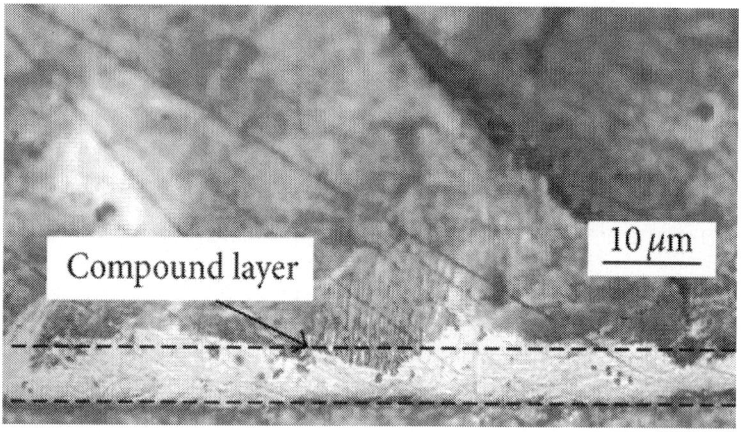

(a)

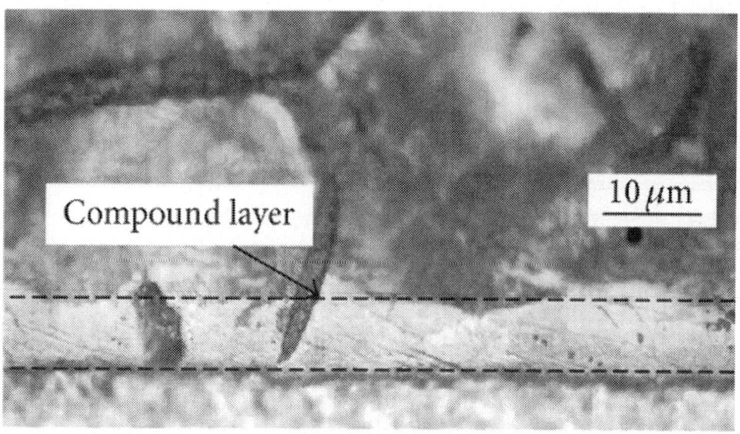

(b)

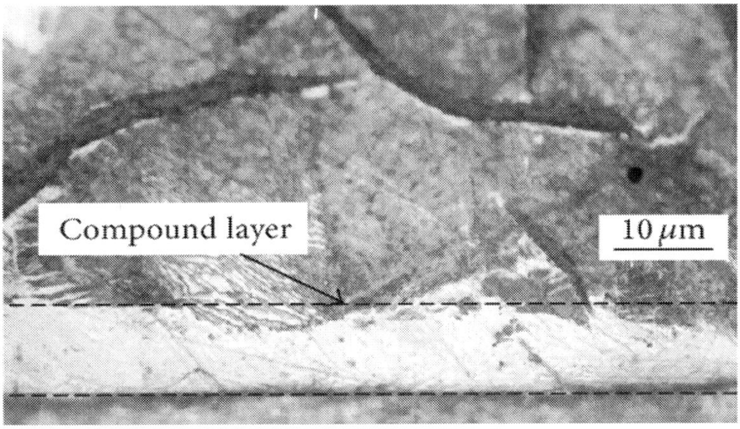

(c)

Figure 2: Cross-sectional scanning electron microscopy observation of the nitrided samples at (a) 550, (b) 570, and (c) 590°C.

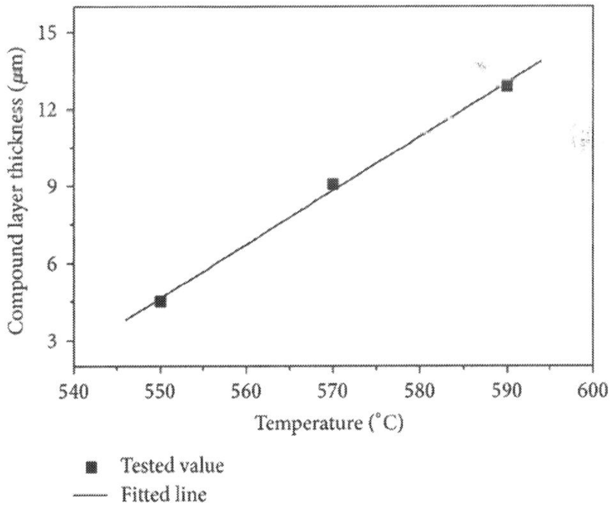

Figure 3: Experimental compound layer thickness as a function of nitriding temperature.

Hardness Measurement

The tested hardness values in the depth of 100 and 130 μm are listed in Table 2. With the increase in distance from the surface, the hardness decreases to some degree. For both 100 and 130 μm, the hardness gets its maximum value at 570°C.

Table 2: The tested hardness of the compound layer at 550, 570, and 590°C

Distance from the surface (μm)	Hardness value (Hv)		
	550	570	590
100	440	501	474
130	403	455	447

The case depth is determined by testing the microhardness of the nitrided samples. The depth is considered to be the distance between the workpiece surface and the location where the hardness is the same with the matrix material. For the analyzed grey cast iron, the obtained average case depths after nitriding are 79, 139, and 155 μm at 550, 570, and 590°C, respectively.

XRD Analysis

Figure 4 presents the XRD diffractogram of the sample's surface. The horizontal and longitudinal coordinate represents the diffractogram angle and diffractogram strength, respectively. Due to the relative thinness of the compound layer, the reflection intensity of the γ'-Fe_4N phase is not distinct enough. Nevertheless, it can be found that after nitriding, the major phases on the surface are γ'-Fe_4N and ε-Fe_3N, which confirms the presence of superficial compound layer.

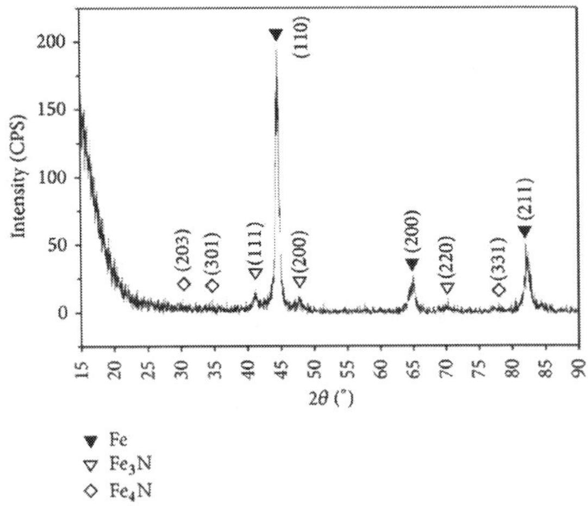

Figure 4: XRD diffractogram of the sample's surface for the analyzed grey cast iron after nitriding.

NUMERICAL RESULTS AND DISCUSSIONS

Numerical Results

The gas nitriding process in the study was simulated by using the model described in Section 3. The calculated nitrogen concentration of grey cast iron is depicted in Figure 5. It can be seen that with the increase in nitriding temperature, the thickness of diffusion zone increases. For a clear presentation, the experimental and calculated thickness values are listed in Table 3 and the calculation errors are also presented. It needs to be noted that the tested thickness of diffusion zone is determined by removing the compound layer from the tested case depth. The calculated thickness values are in good agreement with the experimental data, with an absolute maximal error value of 6.8%, which demonstrates the accuracy and feasibility of the numerical simulations.

Table 3: The calculated and experimental case depths of the grey cast iron at 550, 570, and 590°C

T (°C)	Tested (μm)	Calculated (μm)	Error (%)
550	74	79	6.8
570	129	125	-3.1
590	142	141	-0.7

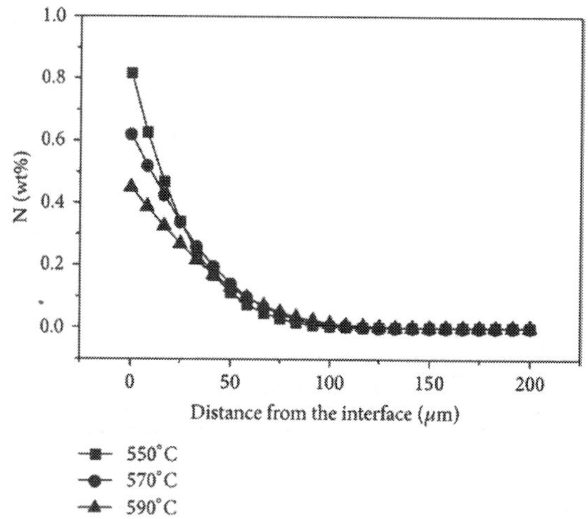

Figure 5: The calculated nitrogen concentration distribution at 550, 570, and 590°C.

For the presence of calculation error, it can be attributed to the following reasons. Firstly, during calculation, the influence of the interface's movement was not considered, which was a disadvantage of the present model. Secondly, the critical value of nitrogen potential, which was directly obtained from the Lehrer diagram of pure iron, may not be accurate for grey cast iron. Furthermore, for the analyzed grey cast iron, the largely distributed graphite may also have some uncertain effects on the nitriding potential value. Nevertheless, the accuracy of the numerical simulations is considered to be acceptable, which is a guarantee of the following effect analysis of nitriding process parameters.

Effect of Nitriding Temperature on the Threshold Value of Nitriding Potential

Nitriding temperature is a crucial process parameter which greatly affects the gas nitriding process. In the study, a threshold value of nitrogen potential is adopted instead of the traditional critical value, which is a significant advantage of the present model in comparison with the existing ones [5–7]. The relationship between the threshold value and temperature is given in Figure 6. As temperature increases, the threshold value decreases significantly. Under the circumstances, the time needed to form the compound layer decreases at higher nitriding temperature.

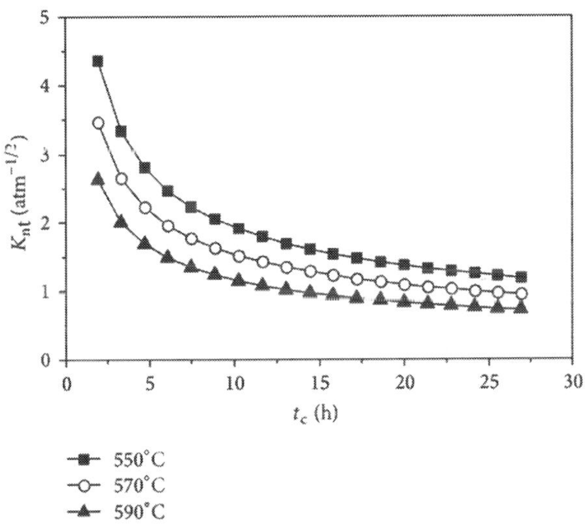

Figure 6: The threshold value of nitrogen potential at different temperatures.

Effect of Nitriding Potential

Nitrogen potential, which represents the nitrogen activity on the workpiece surface, is used to control the compound layer's thickness and phases. In order to investigate the influence of nitrogen potential, numerical simulations were conducted with respect to different nitriding potential values, that is, 0.32, 0.75, 1.5, and 3.5 atm$^{-1/2}$. Figure

7 presents the calculated nitrogen concentration profile for different nitriding potentials. It can be observed that the nitrogen potential has no effect on the case depth. However, the nitrogen concentration on the workpiece surface is of great difference, which exhibits the highest value for nitriding potential of 3.5 atm$^{-1/2}$. The analysis results directly demonstrate that the nitriding potential has great effect on the phase composition of compound layer, since it strongly influences the nitrogen concentration on the surface.

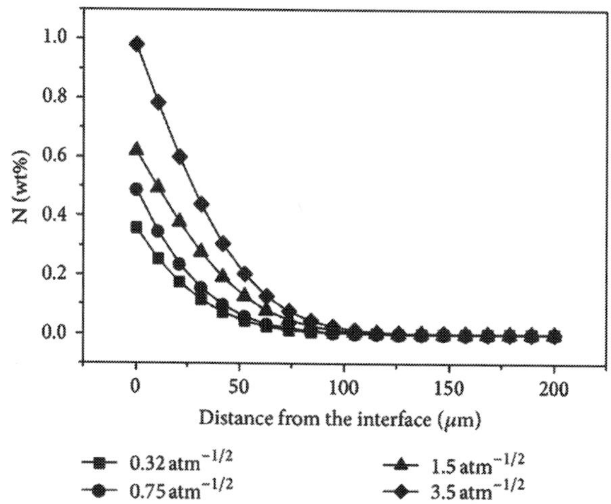

Figure 7: The nitrogen concentration distributions with different nitriding potentials.

Effect of Nitriding Time

To investigate the influence of nitriding time on the case depth, numerical calculations were performed with different nitriding time of 7, 11, 24, and 40 hours. Figure 8 shows the calculated nitrogen concentration profile for different nitriding time. Because the threshold values of nitriding potential at the nitriding time of 7 and 11 hours are higher than those of the present nitriding potential, no compound layer is formed on the surface. When nitriding time is 24 and 40 hours, the compound layer is formed and the nitrogen concentration on the workpiece surface turned into the same value. At the same time, the case depth increases as nitriding time increases from 24 to 40 hours.

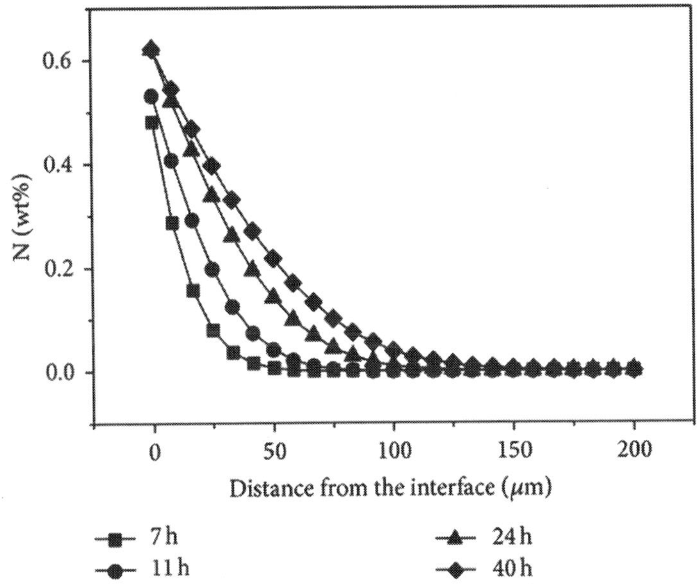

Figure 8: The nitrogen concentration distributions with different nitriding time.

CONCLUSIONS

Based on the gas nitriding experiments and numerical simulations, the following conclusions can be drawn.(1)The numerical simulations, taking into account the influence of threshold value of nitriding potential, can precisely predict the thickness of diffusion zone and nitrogen concentration profile.(2)With the increase in nitriding temperature and time, the thickness of case depth increases to some degree, whereas the nitrogen concentration on the surface decreases as nitriding temperature increases.(3)For the analyzed grey cast iron, the nitriding potential has little effect on the thickness of case depth but has a relatively great influence on the nitrogen concentration of workpiece surface and composition of nitriding layer. The proposed numerical simulations for nitriding layer prediction and analysis can be a reference for practical engineering applications.

REFERENCES

1. M. A. J. Somers, "Thermodynamics, kinetics and microstructural evolution of the compound layer; A comparison of the states of knowledge of nitriding and imitrocarburising," Heat Treatment of Metals, vol. 27, no. 4, pp. 92–102, 2000.
2. D. Pye, Practical Nitriding and Ferritic Nitrocarburizing, Materials Park, Ohio, Ohio , USA, 2003.
3. M. A. J. Somers and E. J. Mittemeijer, "Layer-growth kinetics on gaseous nitriding of pure iron: evaluation of diffusion coefficients for nitrogen in iron nitrides," Metallurgical and Materials Transactions A, vol. 26, no. 1, pp. 57–74, 1995.
4. H. Du, M. A. J. Somers, and J. Agren, "Microstructural and compositional evolution of compound layers during gaseous nitrocarburizing," Metallurgical and Materials Transactions A, vol. 31, no. 1, pp. 195–211, 2000.
5. A. F. M. Arif, S. S. Akhtar, and B. S. Yilbas, "Effect of process variables on gas nitriding of H13 tool steel with controlled nitriding potential," International Journal of Surface Science and Engineering, vol. 4, no. 4–6, pp. 396–415, 2010.
6. M. Yang and R. D. J. Sisson, "Modeling the nitriding process of steels," Advanced Materials and Processes, vol. 170, no. 7, pp. 33–36, 2012.
7. M. Yang, C. Zimmerman, D. Donahue, and J. R. D. Sisson, "Modeling the gas nitriding process of low alloy steels," Journal of Materials Engineering and Performance, vol. 22, no. 7, pp. 1892–1898, 2013.
8. M. J. Hu, J. S. Pan, and Y. J. Li, "Mathematical modelling and computer simulation of nitriding,"Materials Science and Technology, vol. 15, no. 6, pp. 547–550, 2000.
9. M. J. Hu and J. S. Pan, Fundamentals of Thermochemical Surface Treatment of Steel, Shanghai Jiao Tong University Press, Shanghai, China, 1996.
10. W. D. Jentzsch, F. Esser, and S. Boehmer, "Mathematical model for the nitriding of soft iron by ammonia-hydrogen mixtures," Neue Hutte, vol. 26, no. 1, pp. 19–23, 1981.

11. Z. X. Zhu and K. F. Yao, Engineering Materials, Tsinghua University Press, Beijing, China, 2011.
12. L. S. Darken and R. W. Gurry, Physical Chemistry of Metals Istedition London, McGraw-Hiu, 1953.
13. T. Bell, "Controlled nitriding in ammoniahydrogen mixturs," Heat Treatment, vol. 73, pp. 51–57, 1975.
14. Y. Sun and T. Bell, "A numerical model of plasma nitriding of low alloy steels," Materials Science and Engineering A, vol. 224, no. 1-2, pp. 33–47, 1997.
15. T. Spalvins, "Tribological and microstructural characteristics of ion-nitrided steels," Thin Solid Films, vol. 108, no. 2, pp. 157–163, 1983.
16. L. Torchane, P. Bilger, J. Dulcy, and M. Gantois, "Control of iron nitride layers growth kinetics in the binary Fe-N system," Metallurgical and Materials Transactions A, vol. 27, no. 7, pp. 1823–1835, 1996.

Chapter 5

Conducting Polymers / Layered Double Hydroxides Intercalated Nanocomposites

Jairo Tronto[1], Ana Cláudia Bordonal[2], Zeki Naal[3], and João Barros Valim[2]

[1]Universidade Federal de Viçosa - Instituto de Ciências Exatas e Tecnológicas - Campus de Rio Paranaíba - Rio Paranaíba - MG, Brazil

[2]Universidade de São Paulo - Faculdade de Filosofia Ciências e Letras de Ribeirão Preto - Departamento de Química - Ribeirão Preto – SP, Brazil

[3]Universidade de São Paulo - Faculdade de Ciências Farmacêuticas de Ribeirão Preto - Departamento de Física e Química - Ribeirão Preto – SP, Brazil

INTRODUCTION

Layered nanocomposites represent a special class of multifunctional materials that has received a lot of attention over the last years [1-6].

The specific architecture of these composites promotes a synergistic effect between the organic and inorganic parts, generating compounds with different chemical or physical properties as compared with the isolated components. These composites not only represent a creative alternative to the search for new materials, but also allow the development of innovative industrial applications. The potential uses of layered nanocomposites include intelligent membranes and separation devices, photovoltaic devices, fuel cell components, new catalysts, photocatalysts, chemical and biochemical sensors, smart microelectronic devices, micro-optic devices, new cosmetics, sustained release of active molecules, and special materials combining ceramics and polymers, among others [7-18].

A great variety of layered nanocomposites can be prepared from the combination between polymers and layered inorganic solids [1-3]. Compared with the unmodified polymers, the resulting materials present dramatic improvement in properties such as rigidity, chemical and mechanical resistance, density, impermeability to gases, thermal stability, and electrical and thermal conductivity, as well as high degree of optical transparency.

The first successful development concerning the combination of layered inorganic solids with polymers was achieved by researchers from Toyota®, who aimed at structural applications of the nanocomposites in vehicles. These researches prepared nanocomposites by combining nylon-6 and montmorillonite (clay) using the *in situ* polymerization method [20-22]. Research conducted over the past 10 years has shown that nanocomposites containing only a small amount of inorganic silicate (2% volume), exhibit twofold larger elastic modulus and strength without sacrificing resistance to impact. Other automobile companies began to employ this type of material in their vehicles and intensified research in this area [1-3,21].

The excellent gas barrier and vapor transmission properties of these hybrid nanocomposites have led to their application mainly in food industry, more specifically in food and drink packaging. Incorporation of layered silicate nanoparticles into polymeric matrices creates a labyrinth within the structure, which physically retards the passage of gas molecules [22]. These materials can also be used to coat storage tanks in ships and lines of cryogenic fuels in aerospace systems. Compared with the unmodified polymer, nanocomposites delay fire propagation and enhance thermal stability. In contrast to the amount of additives used in traditional fireproof polymers (60%), these nanocomposites contain low layered inorganic solid loading, typically 2-5 wt%. This is due to the formation of an insulating surface layer that not only slows degradation of the polymer, but also decreases its calorific capacity [1,2]. The decomposition temperature of these nanocomposites can be increased to 100 °C, which extends the use of these materials at ambient temperatures, as in the case of automobile engines.

With respect to environmental applications, layered inorganic solids combined with biodegradable polymers have been employed as reinforcing agents. These materials, called "green" nanocomposites, are an attractive alternative for the replacement of petroleum derivatives in the production of plastics.

Depending on the nature of the components and on the preparation method, two main types of nanocomposites can be obtained from the association of layered compound with polymers, as shown in Figure 1:

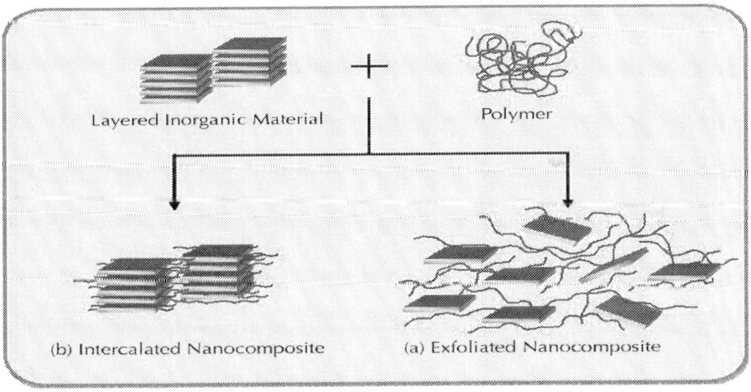

Figure 1: Schematic representation of the different types of composites produced from the interaction between layered compounds and polymers: (a) Intercalated Nanocomposite; (b) Exfoliated Nanocomposite.

- *Intercalated Nanocomposite*: The polymer is intercalated between the inorganic layers, producing a nanocomposite consisting of polymeric chains and alternating inorganic layers. Intercalation of the polymer often results in increased interlayer spacing; *i.e*, larger distance between two adjacent inorganic layers. (Figure 1a).

- *Exfoliated Nanocomposite*: The material presents no ordering along the stacking axis of the layer, or the spacing between the inorganic layers is greater than 8 nm. (Figure 1b).
- In addition to the well-defined structures cited above, a third intermediate type of structure can be found, in which the material presents characteristics of intercalation and exfoliation. In this case, there is broadening of the X ray diffraction peaks.
- Several strategies have been utilized for the preparation of organic-inorganic hybrid materials containing layered inorganic solids and polymer [1,2]:
- *Exfoliation-adsorption*: The layered compound is exfoliated using a solvent in which the polymer is soluble. In some layered compounds there are weak interaction forces between the layers, which can thus be easily exfoliated in appropriate solvents. The polymer may then adsorb onto the exfoliated layers which, after evaporation of the solvent, can be stacked again. As a result, the polymer is intercalated, and an ordered multilayer structure is obtained.
- *In situ intercalative polymerization*: The layered compound undergoes swelling (interlayer expansion) in a solution containing the monomer. The polymer is formed in the interlayer region. The polymerization reaction can be performed by heat or radiation treatment, using an organic initiator or a fixed catalyst.
- *Melted polymer intercalation*: The layered compound is mixed with the polymer matrix in the melting phase. If the layered surfaces are sufficiently compatible with the selected organic polymer under these conditions, the latter penetrates into the interlayer space, generating an intercalated or exfoliated nanocomposite. This technique does not require any solvent.
- *Template Synthesis*: This method can only be used for water-soluble polymers. The layered compound is formed *in situ* in an aqueous solution containing the target polymer on the basis of self-assembly forces, the polymer aids nucleation and growth of inorganic layers. As a result, the polymers are retained between the layers.

Among the inorganic solids used in the preparation of layered nanocomposites, one promising class of material is the Layered Double Hydroxides (LDHs), which have been added to polymers for the synthesis of LDH/polymers nanocomposites [23-26]. LDHs can be structurally described as the stacking of positively charged layers intercalated with hydrated anions [27]. In order to better understand the structure of the LDH, it is appropriate to start from the structure of brucite. In this $Mg(OH)_2$ structure, the magnesium cations are located in the center of octahedra, with hydroxyl groups positioned at their vertices. These octahedra share edges, forming neutral planar layers that are held together by hydrogen bonds. In this type of structure, the isomorphic replacement of bivalent cations with trivalent ones creates a positive residual charge in the layers. For charge balance to be reached in the system, anions should be present between the layers. Together with water molecules, the anions promote stacking of the layers, which culminates in the layered double hydroxide structure displaying a poorly ordered interlayer domain. Not only hydrogen bonding but also electrostatic attraction between the positively charged layers and the interlayer anions hold the layers together in LDHs. A schematic representation of the LDH structure is given in Figure 2.

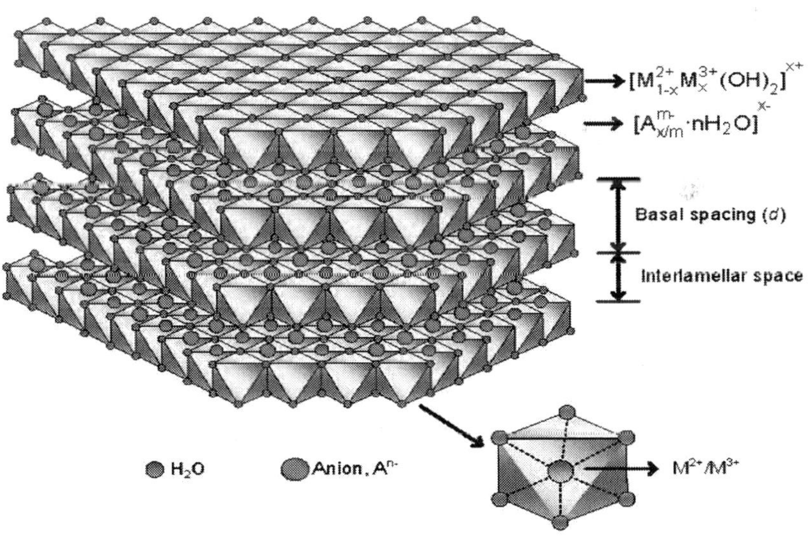

Figure 2: Schematic representation of the LDH structure.

The inorganic layers of the LDH can be stacked according to two different symmetries, resulting in rhombohedral or hexagonal unit cells of the hexagonal system. Most of the synthetic LDH belong to the hexagonal system. Only LDH with M(II)/M(III) ratio equal to 1 are orthorhombic. For the rhombohedral unit cell, the parameter c is three times the basal spacing, space group R3m. In the case of the hexagonal unit cell, the parameter *c* is twice the basal spacing, space group P63mmc. The notations 3R and 2H refer to unit cells as rhombohedral or hexagonal, respectively.

In the structure of LDH, the interlayer domain comprises the region between adjacent inorganic layers. This region is composed of randomly distributed anions and water molecules. Powder X ray diffraction (PXRD) and EXAFS studies, performed by Rousselet et al. showed the highly disordered nature of this region [28]. Besides being found in the interlayer domain, where they hydrate the intercalated anions, the water molecules can also be adsorbed between the crystallites. The water molecules onto the surface of the micro crystallites surface are called extrinsic water molecules, whereas those that are located in the interlayer domain are designated intrinsic water molecules. The global hydration status of the LDH is the addition of both terms, intrinsic hydration and extrinsic hydration. Many researchers consider the interlayer domain of LDH a quasi-liquid state, which gives high mobility to the interlayer anions.

A wide variety of anions can be intercalated into the LDH; for example, organic anions, inorganics and organic-inorganics, and polymers. The intercalation of more than one type of anion in the interlayer domain is an extremely rare phenomenon. Usually, the presence of two or more kinds of anions during the synthesis generates a competition between these anions, and the one with greater tendency to stabilize the system and/or that is present in larger amount will be intercalated. Using PXRD and *in situ* X ray energy dispersion spectroscopy techniques, Fogg et al. reported the existence of a second intermediate stage due to co-intercalation of Cl^- ions and succinate in LiAl-LDH [29]. Pisson et al. studied the exchange of Cl^- anions with succinate and tartarate anions in LDH of the system [Zn_2Al-Cl], [Zn_2Cr-Cl], and [Cu_2Cr-Cl]. The exchange reaction was monitored *in situ* by the X ray diffraction and X ray energy dispersion spectroscopy techniques. The analyses revealed the formation of a second intermediate stage in all the materials, caused by co-intercalation of organic anions and chloride ions [30]. Kaneyoshi and Jones demonstrated that terephthalate anions can adopt two different orientations in relation to the inorganic layers when they are intercalated into Mg-Al-LDH.

The longer molecular axis is either perpendicular or parallel to the plane of the layers. These two orientations are known as interstratified intermediated phases. The occurrence of these two orientations of intercalated terephthalate anions was supported by the appearance of a third basal spacing, attributed to the contribution of two different orientations of anions in the interlayer domain.

A large number of natural and synthetic LDHs containing various metal cations have been studied. In order to form the LDH, the metal cations that will be part of the inorganic layer must present octahedral coordination and ion radius in the range of 0.50 to 0.74 Å. By varying the metal cations, the proportion among them, and the interlayer anion, a large variety of LDH can be prepared. Countless cations can be part of this structure: Mg^{2+}, Al^{3+}, most of the cations of the first transition period, Cd^{2+}, Ga^{3+}, and La^{3+}, among others [27]. In addition LDH displaying more than one bivalent and/or trivalent cation can be synthesized, which further expands the compositional possibilities.

The ratio between the metal cations M(II)/M(III) is also very important, because a change in this ratio between modifies the charge density in the layers, since the charge is generated from the isomorphic substitution of bivalent cations with trivalent ones in the structure of the inorganic layers [27,32]. There is controversy over the values that the x parameter in the general formula of the LDH can assume during the synthesis of these materials. According to de Roy et al., the x value should lie between 0.14 and 0.50, for the formation of an LDH where the M(II)/M(III) ratio can vary between 1 and 6 [33]. For Cavani et al., the x value must fall between 0.20 and 0.34, with the M(II)/M(III) ratio ranging between 2 and 4.37. However, some researchers have reported the synthesis of LDH with different M(II) to M(III) ratios from those mentioned above [27].

As described earlier, the interlayer domain consists of water molecules and anions, mainly. Practically, there is no limitation to the nature of anions that can compensate the residual positive charge of the LDH layers. However, obtaining pure and crystalline materials is not an easy task. Generally, simple inorganic anions with higher charge/radius ratio have greater tendency for intercalation. This is because these anions interact more strongly with the inorganic layers from an electrostatic viewpoint. For the intercalation of organic anions, especially anionic polymers, factors such as the size and geometry of the anion, the interaction between them, and the ratio between size and charge must be taken into account. Some interlayer anions are more mobile, which gives the resulting materials good exchange properties.

Several factors must be borne in mind when planning the synthesis of LDH. For instance, the degree of substitution of M(II) with M(III) cations, the nature of the cation, the nature of the interlayer anion, the pH of the synthesis and, in some cases, the controlled atmosphere. Furthermore, to obtain materials with good crystallinity, the concentrations of the solutions, the rate of the addition of the solutions, the stirring rate, the final pH of the suspension (for variable pH methods), the pH during the addition (for constant pH method), and the temperature of the mixture (typically performed at room temperature) must be controlled. There are a number of methods that can be used for the synthesis of LDH. They can be divided into two categories:

- Direct synthesis methods: salt-base method or co-precipitation (at variable pH or at constant pH), salt-oxide method, hydrothermal synthesis, induced hydrolysis, sol-gel method, and electrochemical preparation [27,33-37].
- Indirect synthesis methods: simple anionic exchange method, anionic exchange by regeneration of the calcined material, and anion exchange using double phase, with formation of a salt between the surfactants [33,38,39].

Among the most extensively investigated conducting polymers are polyacetylene; poly-heterocyclic five-membered compounds like polypyrrole, polythiophene, and polyfuran; and polyaromatics such as polyaniline and poly (p-phenylene).The structures and respective electrical conductivity values of some conductive polymers are summarized in Table 1 [40].

Table 1: Structure and electrical conductivity values of some conductive polymers [40]

Name	Structural Formula	Conductivity / S.cm-1
Polyacetylene		103 a 106
Polyaniline		10 a 103
Polypyrrole		600

Poly(p-phenylene)		500
Polythiophene		200

Below, we will outline some classes of conducting polymers:

Polyaniline

Polyanilines are widely studied because of their low cost, good stability in the presence of oxygen and water, and interesting redox properties. In 1835, polyaniline was first synthesized as "black aniline", a term used for the product obtained by oxidation of aniline [41]. Some years later, Fritzche analyzed the products obtained by chemical oxidation of this aromatic amine [42]. In 1862, Letheby found that the anodic oxidation of aniline in a platinum electrode, in an aqueous solution of sulfuric acid, formed a dark brown precipitate [43]. The polyaniline chain consists of units present in two main forms: (i) the fully reduced form, which contains only aromatic rings and nitrogen atoms of the amine function, shown in Figure 3a, and (ii) a completely oxidized form displaying iminic nitrogen atoms, quinonics rings, and aromatic rings, as represented in Figure 3b.

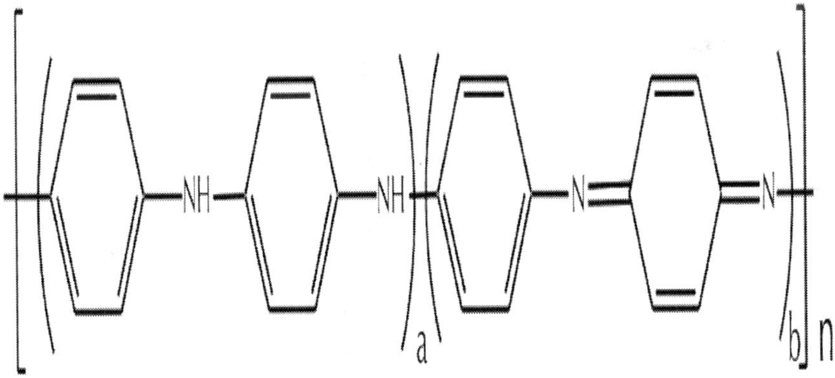

Figure 3: Representation of the general structure of the polyaniline base form: (a) reduced species (b) oxidized species.

Gospodinova and Terlemezyan examined the oxidation state of polyaniline constituents. The principal oxidation states of polyaniline are presented in Table 2 [44]. The emeraldine salt is the structural form of polyaniline that affords higher conductivity values. Polyaniline can be doped by protonation, with no change in the number of electrons in the polymer chain.

Table 2: Most important oxidation states of polyaniline [44]*

Oxidation state	Strucuture	Color	Characteristic
Leucoemeraldine		Yellow 310	Insulating, completely reduced
Emeraldine salt		Green 320, 420, 800	Conductive, half-oxidized
Emeraldine base		Blue 320, 620	Insulating, half-oxidized
Pernigraniline		Purple 320, 530	Insulating, completely oxidized

[i] - * The numerical values refer to the wavelength (in nanometers) where absorption is maximum.

Polypyrrole

The first report about pyrrole was published by Runge in 1834. This author observed a red component in coal tar and bone oil. The compound isolated and purified from this component was named of pyrrole [45]. The structural formula of pyrrole was established in 1870. In the late 19th century, the interest in pyrrole and its derivatives was aroused, following the discovery that this molecule was part of some porphyrins found in biological systems, such as chlorophyll. Approximately 100 years after the first report on the discovery of pyrrole, *i.e*, in 1970, interest in these materials increased again due to the possibility of preparing conducting polypyrroles [46].

Pyrrole is a five-membered cyclic compound (heterocyclic) containing 6 ϖ electrons. Additionally, pyrrole has an sp^2 nitrogen, and its three bonds are located in the plane of the ring. The excess electron, the conjugation of the double bonds, and their ability to relocate are the structural characteristics underlying the charge conduction properties presented by the pyrrole polymer products.

Polythiophene

In 1882, Meyer discovered thiophene [47]. At that time, his studies revealed that this compound, which was isolated from benzene impurities, was a new aromatic system. Thiophene is not a component of animal metabolism, but some thiophene derivatives can be found in plants. Thiophene derivatives are widely employed in many types of chemical industry, including the pharmaceutical, veterinary, polymers, and agrochemicals industries.

Thiophene is a compound analogous to pyrrole. Instead of the nitrogen heteroatom, it contains a sulfur atom with sp^2 hybridization. The sp^2 orbital, which is perpendicular to the ϖ electron system, has an unshared electron pair. The p-orbital of sulfur donates two electrons to the ϖ system. Polythiophene derivatives have been extensively studied, probably because most of them are soluble in organic solvents, which facilitates processing of the material.

The electrical conductivity of a solid is the result of the number of charge carriers (electrons / holes) and their mobility. Conducting polymers have a large number of charge carriers with low mobility, which is mainly caused by the large number of structural defects such as reticulation and the disordering of chains. The formation of nanocomposites by intercalation of conductive polymers into LDH can minimize the formation of reticulation defects and the disordering of polymer chains, furnishing materials with new and interesting properties. Table 3 summarizes the literature works on the synthesis and characterization of conducting polymers intercalated into LDH.

Table 3: Some examples of Nanocomposites consisting of LDH/conductive polymers

Year	Nanocomposites	Examples	Authors	Ref.
1994	LDH / polyaniline	CuCr- polyaniline-LDH CuAl- polyaniline-LDH	Challier and Slade	48

2001	LDH / aminobenzoate derivatives	LiAl-o-, p- and m-aminobenzoate-LDHs	Isupov et al.	49
2002	LDH / aniline sulfonate	CuCr- aniline sulfonate-LDH	Moujahid et al.	50
2003	LDH / aminobenzene sulfonate	CuCr- aminobenzene sulfonate-LDH	Moujahid et al.	51
2004	LDH / 2-thiophenecarboxylate	ZnAl-2-thiophenecarboxylate-LDH ZnCr-2-thiophenecarboxylate-LDH	Tronto et al.	52
2005	LDH / aniline sulfonate derivatives	CuCr-o- and m-aminobenzenesulfonate, 3-amino-4-methoxybenzenesulfonate, 3-aniline-1-propane sulfonate, and 4-aniline-1-butane sulfonate-LDHs.	Moujahid et al.	53
2006	LDH / pyrrol derivatives	ZnAl-4-(1H-pyrrol-1-yl) benzoate-LDH ZnCr-4-(1H-pyrrol-1-yl) benzoate-LDH ZnAl-3-(pyrrol-1-yl)-propanoate-LDH ZnAl-7-(pyrrol-1-yl)-heptanoate-LDH	Tronto et al.	54 55 56
2006	LDH / aminobenzoate derivatives	MgAl-aminobenzoate-LDH NiAl-aminobenzoate-LDH	Tian et al.	57 58
2006	LDH / aniline sulfonic	NiAl-aniline sulfonic -LDH	Wei et al.	59

Challier and Slade reported the synthesis and characterization of layered nanocomposites of CuCr and CuAl-LDHs intercalated with polyaniline [48]. The oxidizing host matrices were prepared by the coprecipitation method, with the intercalation of terephthalate anions into CuCr-LDH and hexacyanoferrate(II) anions into CuAl-LDH. Then, the LDH precursors were submitted to an anion exchange reaction with a solution of pure aniline under reflux, for 24 h. The X ray diffractograms showed that the materials submitted to reaction with aniline exhibited basal spacings of 13.3 Å and 13.5 Å for CuCr-LDH and CuAl-LDH, respectively. This result was consistent with the intercalation of aniline molecules containing aromatic rings oriented perpendicular to the plane of the layers. FTIR analyses evidenced polymerization of the aniline molecules, since the absorption spectra displayed bands typical of the emeraldine form. According to the authors, the oxidant character of Cu^{2+}, present in layered structure of the inorganic host, helped induce oxidative polymerization of the aniline intercalated in the interlayer galleries.

Isupov et al. described the intercalation of *o*-, *p*-, and *m*-aminobenzoate anions into LiAl-LDH [49]. The incorporation of aminobenzoate anions in the host matrices was conducted by anion exchange, from an LiAl-LDH intercalated with chloride anions. The basal spacings obtained from the X ray diffractograms indicated that the anion exchange reaction was effective, with incorporation of aminobenzoate anions in the interlayer domain. To carry out the *in situ* polymerization, samples of the nanocomposites were submitted to a heat-treatment at 90 °C for 100 h, with 75% relative air humidity. For the LiAl-LDH intercalated with *m*-aminobenzoate anions, the formation of a polyconjugated system was confirmed by ESR spectra performed in vacuum at 77 K and 300 K. The spectra displayed a broad isotropic signal between 7.3 and 7.5 G, with g = 2.000 and line with Gaussian shape. Heating of the nanocomposite in air intensified the ESR signal. Formation of the polyconjugated system was also corroborated by FTIR and Raman spectroscopies. A comparison of the FTIR spectrum of LiAl-LDH intercalated with *m*-aminobenzoate and its oxidation products revealed a marked decrease in the band located at 1250 cm^{-1}, which indicates a decrease in the number of amino groups, $-NH_2$.

Moujahid et al. reported the intercalation of the *m*- and *o*-aminebenzeno sulfonate anion, 3-amine-4-methoxybenzene sulfonate, 3-aniline-1-propane sulfonate and 4-aniline-1-butane sulfonate into Cu_2Cr-LDHs [50,51,53]. These authors discussed the arrangement of intercalated molecules and their subsequent dimerization and/or *in situ* polymerization. The authors incorporated these inorganic anions between the layers was by the direct precipitation method at constant pH. After the synthesis, the resulting materials were submitted to heat treatment at different temperatures, under air atmosphere. The interlayer distances of they synthesize nanocomposites were consistent with the presence of a bilayer of guest molecules in the interlayer space. The heat treatment performed at 350 K culminated in a contraction in the basal spacing of the nanocomposites, except for the material intercalated with the *o*-aminebenzene sulfonate anions. This contraction was associated with reorientation of the intercalated molecules and/or with an *in situ* polymerization. At temperatures above 350 K up to approximately 450 K, there was no significant variation in basal spacing. ESR analysis evidenced *in situ* polymerization. The profiles of the ESR spectra changed with increasing temperature. The value of the signal g ($g = 2.0034 \pm 0.0004$) was typical of the formation of organic radicals and/or conduction electron. For the nanocomposite intercalated with the *m*-aminobenzene sulfonate anion, the ESR and CV analysis showed that the *in situ* polymerization must occur with a syndiotactic arrangement. For the nanocomposite intercalated with the *o*-aminobenzene sulfonate anion, the ESR studies indicated a very weak response of spin carriers for these materials. The electrochemical characterization did not show the presence of reversible redox processes. These results, together with the constancy of the basal spacing value obtained up to a temperature of 450 K, suggested that *in situ* polymerization was not favored when the monomer had the amino group located at the *ortho* position relative to the sulfonate group. For the 3-amine-4-methoxybenzene sulfonate anion, the presence of methoxy group in the *para*position relative to the sulfonate group made the polymerization process difficult. The ESR and CV data for this nanocomposite indicated formation of a dimer. For the nanocomposites synthesized with 3-aniline-1-propane sulfonate and 4-aniline-1-butane sulfonate, the heat-treatment at 473 K prompted an increase in g ($g = 2.0034 \pm 0.0004$), which is associated with the generation of organic radicals and/or conduction electrons.

Tronto et al. described the synthesis, characterization, and electrochemical investigation of 2-thiophenecarboxylate intercalated into ZnAl-LDH and ZnCr-LDH [52]. The materials were synthesized by the coprecipitation method at constant pH, followed by hydrothermal treatment at 65 °C for 72 h. The LDH were analyzed by PXRD, FT-IR, ^{13}C CP-MAS, TEM and CV. The basal spacing was about 15.3 Å for all the LDH which suggested the formation of bilayers of anions intercalated between the inorganic sheets. In this configuration, the 2-thiophenecarboxylate anions would be in a position in which their longer axes would lie perpendicular to the plane of the inorganic layers. Besides the phase with basal spacing of 15.3 Å, another phase with basal spacing of 7.58 Å was also detected in the diffractograms. This value was similar to some values reported for the intercalation of CO_3^{2-} anions into ZnAl-LDH and ZnCr-LDH. However, the qualitative analysis and ^{13}C CP-MAS did not confirm the presence of carbonate anions, as contaminant in the LDH. Thus, the results indicated that for this second phase, 2-thiophenecarboxylate anions were intercalated with their longer axes parallel to the plane of the inorganic layer. ^{13}C CP-MAS data further suggested that, during the synthesis, the 2-thiophenecarboxylate anions lost an acid hydrogen which led to formation of the dimer.

Tronto et al. conducted a study on the *in situ* polymerization of pyrrole derivatives, 4-(1H-pyrrol-1-yl)benzoate, 3-(Pyrrol-1-yl)-propanoate, and 7-(pyrrol-1-yl)-heptanoate, intercalated into LDH [54-56]. The materials were synthesized by co-precipitation at constant pH, followed of hydrothermal treatment for 72 h. The final LDH were characterized by X ray diffraction, ^{13}C CP-MAS NMR, TGA, and ESR. The basal spacing value coincided with the formation of bilayers of intercalated monomers. ^{13}C CP-MAS NMR and ESR analyses showed the formation of a polyconjugated system with polymerization of the monomers intercalated in the LDH during the coprecipitation and/or hydrothermal treatment processes. This result reinforced the authors assumption that the connectivity between the monomers occurred spontaneously during the synthesis, with generation of oligomers and/or syndiotactic polymers intercalated between the LDH layers. At room temperature, the ESR spectrum displayed a signal typical of the hyperfine structure (hfs). The presence of hfs suggested the existence of a proper regulatory environment for the free electrons. These electrons would be present in an organic "backbone" of small size. Thermal analysis of these materials revealed that the inorganic host matrix provided the intercalated polymers with thermal protection, because the thermal decomposition reactions happened at higher temperatures compared with the pure polymers.

Tian and cols. investigated the oxidative polymerization of *m*-$NH_2C_6H_4SO_3^-$ anions intercalated into NiAl-LDH using ammonium persulfate as the oxidizing agent [57]. The amount of oxidizing agent required for controlled polymerization of the intercalated monomers was systematically evaluated. The materials were characterized by PXRD, UV-Vis spectroscopy, FT-IR spectroscopy, and XPS determination. PXRD and elemental analysis data showed the co-intercalation of nitrate anions, originating from the LDH precursor, and *m*-$NH_2C_6H_4SO_3^-$ anions. UV-Vis results evidenced polymerization of the intercalated *m*-$NH_2C_6H_4SO_3^-$ anions, with the formation of small chains. The intercalated polyaniline sulfonate was present in different oxidation states and at different protonation levels, depending upon the amount of oxidizing agent that was added.

Tian and cols. also performed the *in situ* oxidative polymerization of *m*-$NH_2C_6H_4SO_3^-$ anions intercalated into MgAl-LDH [58]. The monomers were incorporated into the LDH via an exchange reaction using the precursor $[MgAl(OH)_6](NO_3)$ nH_2O. The nitrate anions remaing from the exchange reaction and co-intercalated with the *m*-$NH_2C_6H_4SO_3^-$ anions were utilized as oxidizing agent for the oxidative polymerization of the intercalated monomers. The resulting materials were analyzed by DTA-TG-DSC as well as UV-Vis and HT-XRD spectroscopies. In the temperature range 300-350 °C, the UV-Vis analysis confirmed the reduction of nitrate and polymerization of aniline.

Wei et al. reported the oxidative polymerization of *m*-$NH_2C_6H_4SO_3^-$ anions in NiAl-LDH, using intercalated nitrate anions as the oxidizing agent [59]. The LDH interlayer space was used as a "nanoreactor" for the *in situ* polymerization of the intercalated monomer. Polymerization of the monomers was acomplished by heat treatment under nitrogen atmosphere. The interlayer polymerization was monitored by thermogravimetric analysis coupled with differential thermal analysis and mass spectrometry (TGA-DTA-MS), UV-Vis spectroscopy, X ray absorption near edge (XANES), (HT-XRD) and FTIR spectroscopy. Polymerization of the monomer was observed at a temperature of 300°C.

SYNTHETIC STRATEGIES FOR THE PREPARATION OF CONDUCTING POLYMERS / LAYERED DOUBLE HYDROXIDES INTERCALATED NANOCOMPOSITES

The synthesis of intercalated nanocomposites of LDH/conductive polymers can be carried out using different strategies. The main ones are [23]:

- Intercalation of monomer molecules between the LDH layers, with subsequent *in situ*polymerization. Intercalation of the monomer can occur by direct or indirect methods. The intercalation of monomer molecules with subsequent *in situ* polymerization, is widely used in the preparation of various LDH/conductive polymers. The resulting nanocomposites generally exhibit good structural organization and phase purity. This process is limited by two factors:
  1. the distance between the monomers when they are strongly linked, or grafted, to the structure of the inorganic layers. When the monomers are strongly bound to the layers, their flexibility (freedom of movement within the interlayer) is limited, so the proximity between them should be sufficient for the polymerization reaction to occur. High charge densities in the layers may shorten distance between the intercalated monomers. Functionalized monomers with long chain aliphatic groups also provide greater flexibility.
  2. the polymerization conditions (temperature, pH, or redox reaction), which should be selected so as not to affect the layered structure of the resulting materials.

Indirect methods may also be employed for the intercalation of monomers. These methods are often utilized when the chemical nature of the interlayer space and guest species are not compatible. Such methods require the preparation of an LDH precursor intercalated with a molecule that can be easily exchanged. This LDH precursor is then placed in contact with the monomer of interest, which will replace the previously intercalated anion. To obtain the LDH/polymer, it is necessary to carry out the *in situ* polymerization reaction after the exchange with the monomer.

- Direct intercalation of polymer molecules with low molecular weight between the LDH layers or intercalation polymers with high molecular weight by indirect methods. The incorporation of the polymer between the LDH layers, can be performed by direct method by using the direct co-precipitation reaction, nanocomposites containing polymers that have an anionic group; for example, carboxylate or sulfonate groups, can be produced during growth of the inorganic crystal. This preparation strategy usually yields nanocomposites with low structural organization. The crystallinity of these materials can be improved by hydrothermal treatment. The indirect method requires the presence of the LDH precursor, usually containing chloride anions. This LDH precursor is placed into exchange reaction using suitable solvents in the presence of the polymer of interest.
- Intercalation of LDH via exfoliation, when a colloidal system is formed between the LDH and an appropriate solvent, for exfoliation of the layers. Restacking of the layers in the presence of a solution containing the target monomer or polymer culminates in their intercalation by restacking of the structure of the layer. When the monomers are intercalated, a subsequent *in situ*polymerization is required for attainment of the intercalated nanocomposite LDH/polymer. This strategy is usually employed when the polymer has high molecular weight, which makes their diffusion between the LDH layers difficult. Due to its high charge density, the LDH does not have a natural tendency to exfoliation. To achieve delamination of these materials, it is necessary to reduce the electrostatic interaction between the layers. This can be done with intercalation of spacer anions, such as, dodecylsulfonate and dodecylbenzenesulfonate. Exfoliation is then obtained by placing the organically modified LDH in a solution containing a polar solvent. Addition of polymer to the solution containing the exfoliated material results in the formation of an intercalated and/or exfoliated precipitate. In some cases, the nanocomposite is only generated upon evaporation of the solvent.

In addition to the strategies described above, immobilization of the polymer between the LDH inorganic sheets can also be attained by regeneration of the layered structure using the *"memory effect"* exhibited by some LDH. In this case, a previously prepared LDH, normally MgAl-CO_3, is firstly calcined at an adequate temperature, for elimination of the interlayer anion. The calcined material, a mixed oxy-hydroxide, is then placed in contact with an aqueous solution of the polymer to be intercalated. The oxide is hydrolyzed with regeneration of the LDH structure and intercalation of the polymer. This process is accompanied by a sharp increase in the pH value. The latter can be corrected, to prevent the intercalation of hydroxyl anions. Normally, the LDH/polymers nanocomposites produced by this method do not exhibit good organization, being more suitable for the incorporation of small molecules. This method was used for the intercalation of silicates into LDH. In this case, mexinerite (an MgxAlOH-LDH, with x = 2, 3, 4) was employed as precursor for incorporation of the silicate to this end, mexinerite was previously calcined at 500 °C under air atmosphere, and then placed in contact with a solution of tetraethylorthosilicate, $Si(OC_2H_5)_4$ (TEOS). This afforded more crystalline materials than those obtained by anion exchange or direct co-precipitation, using metasilicate and ZnM-LDHs (M = Al, Cr).

An additional route for preparation of the LDH/polymer is the auxiliary solvent method. Solvents represent an important part in the swelling processes of the layered materials, since they promote separation of the layers. Schematic representation of the incorporation of polymers into layered double hydroxides is given in Figure 4.

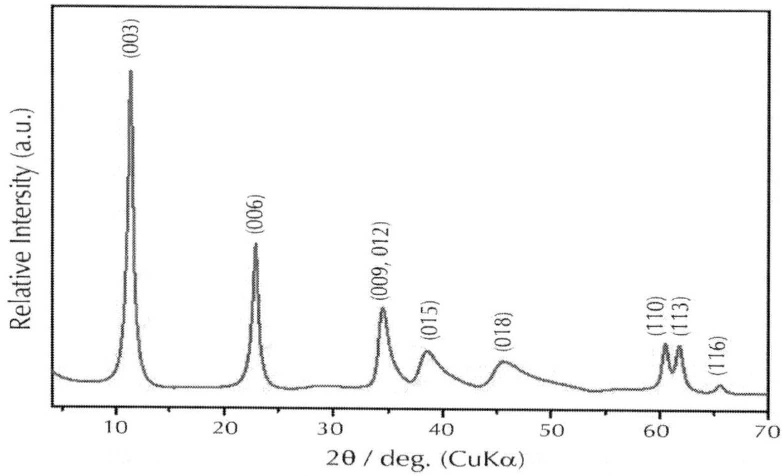

Figure 4: Schematic representation of the incorporation of polymers into layered double hydroxides. *(adapted from ref. 23)*

CHARACTERIZATION METHODS

This section describes the main techniques employed for the characterization of intercalated nanocomposites of conducting polymer / LDH: Powder X-ray Diffraction (PXRD), ^{13}C Cross-Polarization/ Magnetic Angle Spinning (CP/MAS) NMR spectroscopy, Electron Spin Resonance (ESR) spectroscopy, Thermogravimetric Analysis (TGA), Differential Scanning Calorimetry (DSC), Fourier Transform Infrared (FTIR) spectroscopy, Ultraviolet/Visible (UV-Vis) Spectroscopy, Transmission Electron Microscopy (TEM), and Scanning Electron Microscopy (SEM).

Powder X-Ray Diffraction (PXRD)

The X ray diffraction pattern (PXRD) of LDH presents basal peaks 00*l* related to the stacking sequence of the inorganic sheet. The peaks are not basal, said to non harmonics, are related to the sheet structure. For new LDHs, the indexing of the diffraction peaks can be accomplished by comparison with the PXRD of hydrotalcite, which exists in the database of diffraction equipment (JCPDS-ICDD, PDF database), or with a number of other LDHs described in the literature. Figure 5 brings a representative PXRD for an MgAl-CO_3-LDH.

The interlayer distances can be calculated from the values of 2θ, using the Bragg equation:

$$n\lambda = 2d_{hkl} \cdot sen\theta$$

where n is the diffraction order, d_{hkl} is the interlayer spacing for the peak hkl, and θ is the Bragg angle, determined by the diffraction peak. Repetition of the d value, for $n = 1, 2, 3...$, evidences the formation of a layered material. The interlayer spacing can be calculated by averaging the basal peaks according to the equation:

$$d=\frac{1}{n}(d_{003}+2d_{006}+...+nd_{00n})$$

The parameters a and c can be obtained according to the equation:

$$\frac{1}{(d_{hkl})^2}=\frac{4}{3}\left(\frac{h^2+hk+k^2}{a^2}\right)+\frac{l^2}{c^2}$$

where h, k, and l are the Miller indices of the corresponding peak. For a LDH with stacking sequence 3R, the c parameter c is equal to three times the basal spacing value.

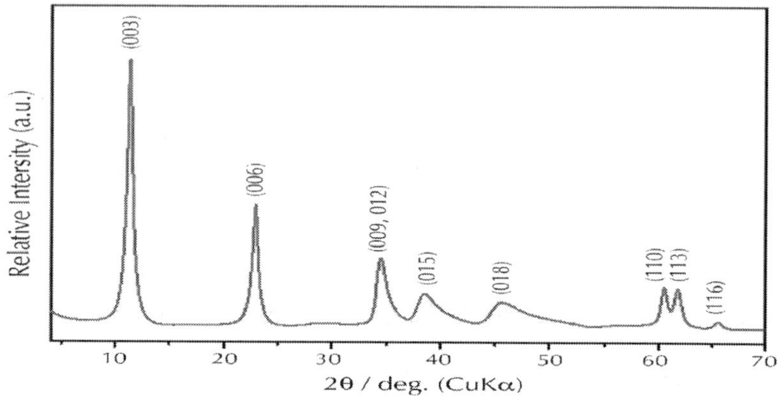

Figure 5: PXRD of synthetic Hydrotalcite.

To determine the orientation adopted by anionic species, such as monomers and polymers intercalated into LDHs, the values of interlayer spacing and/or basal spacing obtained from the PXRD data are compared with the size of anions obtained by specific computer programs, like "VASP (*Vienna Ab-initio Simulation Package*)".

When thermal treatments is performed for the *in situ* polymerization of monomers intercalated between the LDHs inorganic layers, the PXRD analysis may reveal a decrease in the value of interlayer spacing, which indicates a small contraction between adjacent layers. The presence of phases, other than the LDH can also be identified by PXRD, which is useful since thermal treatment may often generated oxides.

^{13}C Cross-Polarization/Magnetic Angle Spinning (CP/MAS) NMR Spectroscopy

In situ polymerization of monomers intercalated into LDH may be monitored by ^{13}C Cross-Polarization/Magnetic Angle Spinning (CP/MAS) NMR spectroscopy. This technique detects formation of bonds of the monomer-monomer type in polyconjugated systems. Assignment of the chemical shift values for the monomers can be carried out by computer simulation using specific computer programs, such as "*ACD/ChemSketch, version 4.04*", provided by the company Advanced Chemistry Development Inc., and "*CS Chemdraw Ultra®*", offered by the company Cambridgesoft Corporation. The values obtained by simulation can be compared with the values achieved experimentally.

Figure 6 contains an example of ^{13}C Cross-Polarization/Magnetic Angle Spinning (CP/MAS) NMR Spectroscopy analyses for the *in situ* polymerization of 4-(1H-pyrrol-1-yl)benzoate intercalated into ZnAl-LDH [54]. Assignment of the peaks to the carbons of the monomer is given in Table 4.

Table 4: Assignment of the peaks to the carbons in the ^{13}C CP-MAS NMR analyses of 4-(1H-pyrrol-1-yl)benzoate

Notation	Assignment (Cn)	DMSO-D6 (ppm)	CP-MAS (75.4 MHz) (ppm)
	C1	111.3	111.2
	C2	118.9	120.1
	C3	143.0	142.3
	C4	118.5	115.6
	C5	130.9	132.9
	C6	127.0	122.8
	C7	166.6	174.9

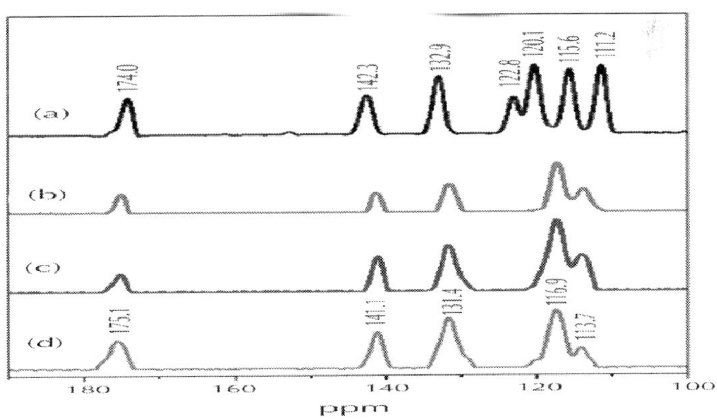

Figure 6: CP-MAS NMR spectra of: (a) pyrrolebenzoic acid; (b) ZnAl-4-(1H-pyrrol-1-yl)benzoate-LDH; (c) ZnAl-4-(1H-pyrrol-1-yl)benzoate-LDH with hydrothermal treatment; and (d) ZnAl-4-(1H-pyrrol-1-yl)benzoate-LDH after thermal treatment [54].

The ^{13}C CP/MAS NMR spectra of all LDH were similar. In the Figure 6c and 6d, the peaks can be unambiguously assigned as carbons C7 (175.1 ppm), C5 (131.6 ppm), C3 (140.8 ppm), and C1 (113.7 ppm). The broad signal at 116.8 ppm can be attributed to the chemical shifts of the remaining carbons C4 and C6 of the six-membered ring. Several simulations of the ^{13}C NMR spectra suggest that one possible quaternary carbon, resulting from the polymerization of the monomer via condensation C2-C2, presents chemical shift in the range of 112.0 to 116.0 ppm. Therefore, the large signal at 116.8 ppm in spectrum of the polymer is ascribed to this quaternary carbon, coinciding with the chemical shifts of the remaining carbons C4 and C6. Together with the PXRD results, these data suggest that the production of oligomers and/or polymers occurs with the formation of bilayers of monomers in the interlayer space. In this arrangement, the carboxylate groups are directed to the layer, whereas the aromatic rings occupy the central region of the interlayer spacing. Therefore, the polymer obtained within the interlayer resembles a *"zig-zag"*, similar to the polymers of the syndiotactic type.

Electron Spin Resonance (ESR) Spectroscopy

Electron Spin Resonance (ESR) spectroscopy allows for monitoring of *in situ* polymerization processes in intercalated monomers. The spectra of conducting polymers usually exhibit signs typical of the formation of polarons, with the Lorentzian profile. In these analyses, the apparatus is normally operated at 9.658 GHz, using the 1,1-diphenyl-2-picrylhydrazyl (DPPH) radical to determine the ressonance frequency ($g = 2.0036 +/- 0.0002$). The scan width can vary between 2000 and 4000 G, with a receiver gain of 100000.

Figure 7 illustrates ESR analyses for monitoring of the *in situ* polymerization of 3-(Pyrrol-1-yl)-propanoate monomers intercalated into ZnAl-LDH [55, 56]. The spectra were recorded after heat-treatment at temperatures ranging from ambient to 180 °C for 2h. For material at room temperature, the ESR spectra display very weak signals. Thus, the spectrum was enlarged 16 times for comparison with those of the material treated at other temperatures. Typical signs can be noticed for the *"superhyperfine"* structure with 6 lines, and there is a sign characteristic of the formation of a polaron with $g = 2.004 \pm 0.0004$. The appearance of this *"superhyperfine"* structure suggest formation of the radical (COO·). The magnetic moment of this radical should interact with the magnetic moments of the nuclei of the aluminum atoms present in the inorganic host matrix. This hypothesis considers the nuclear spin of aluminum as I = 5/2 and a number of nuclei N = 1, which generates a spectrum of 2NI + 1 = 6 lines. Due to the charge balance required for maintenance of the electroneutrality of hybrid systems there a regulating environment for the free electrons of the radicals (COO ·) within the interlayer spacing. These radicals are located near the aluminum cations, because the latter are responsible for the positive charge density of the layer. The signal at g = 2.004 ± 0.0004 attests to the formation of a polaron, *i.e.*, a polarized entity resulting from delocalization of the radical in structures with ϖ conjugations. The increase in the delocalization of ϖ orbitals favors the generation of polarons, so an increase signal in this upon heat-treatment indicates stronger connection between the monomers. The Lorentzian profile of the ESR spectrum of the material at room temperature is compatible with the formation of conjugated polymers. The ESR results agree with the NMR results and indicate that spontaneous partial polymerization and/or oligomerization of the 3-(Pyrrol-1-yl)-propanoate monomers takes place during coprecipitation of the nanocomposites.

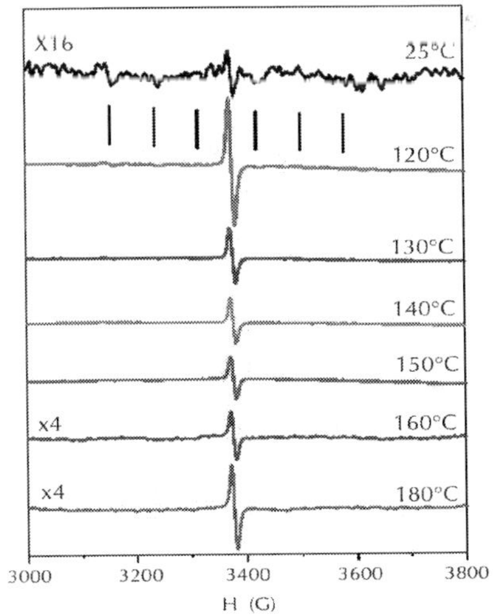

Figure 7: ESR spectra of Zn-Al-3-(Pyrrol-1-yl)-propanoate-LDH as a function of the heat-treatment temperature [55,56].

Thermogravimetric Analysis (TGA) and Differential Scanning Calorimetry (DSC)

The thermal stability of LDHs intercalated with conductive polymers as well as the amount of water, intercalated and adsorbed, in the nanocomposites can be determined by thermogravimetric analysis. The results are obtained as a curve mass decrease (%) *versus* temperature. For the LDHs, the thermal decomposition steps generally overlap, especially in the case of LDHs intercalated with organic molecules.

TGA is important in the thermal *in situ* polymerization of nanocomposites, since it is necessary to determine the temperature that should be used for polymerization of the intercalated monomers. The thermal decomposition of intercalated organic compounds takes place at higher temperatures, so it is possible to achieve greater thermal stability for conducting polymers intercalated into an inorganic host matrix (LDH).

Figure 8 displays an example of TGA/DSC analysis for the ZnAl-LDH intercalated with 3-aminobenzoate monomers and for the pure monomer.

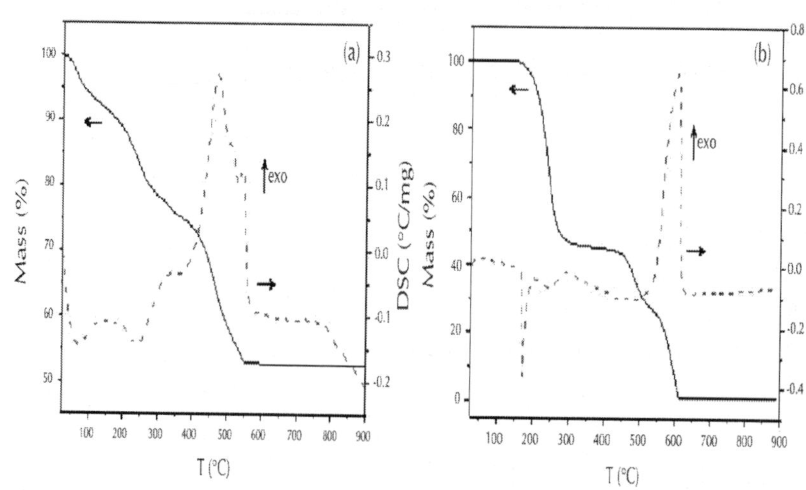

Figure 8: TGA-DSC of (a) ZnAl-3-aminobenzoate-LDH; (b) sodium 3-amino-benzoate [60].

For the nanocomposite material, Figure 8a, the early stages of thermal decomposition are associated with loss of adsorbed and intercalated water. In this temperature range, the DSC curve reveals the occurrence of endothermic processes. Dehydroxylation of the inorganic sheets and decomposition of the anion intercalated species happen concomitantly. The DSC curve also indicates the occurrence of an exothermic process during decomposition of the intercalated organic species.

Fourier Transform Infrared (FTIR) Spectroscopy

FTIR analysis is carried out in KBr pellets, pressed from a mixture of 2% of the LDH samples in previously dried KBr. The spectra are recorded over a wavelength range going from 4000 to 400 cm^{-1}. FTIR spectroscopy data provide information about the functional groups and possible interactions between the organic and inorganic parts of the nanocomposites. Identification of the *in situ*polymerization of monomers intercalated into LDH by this technique is difficult because of several overlapping spectral bands.

Figure 9 contains the FTIR spectra of (a) sodium 3-aminobenzoate, (b) pure sodium poly-3-aminobenzoate, and LDH intercalated with sodium 3-aminobenzoate submitted to different treatments [61]. For the pure monomer, the bands (not shown in the figure) at 3408, 3349, and 3223 cm^{-1} are related to $\nu(NH_2)$ symmetric and anti-symmetric stretching, whereas the band at 1628 cm^{-1} is characteristic of $\delta(NH_2)$ symmetric deformation. The bands at 1560 and 1411 cm^{-1} are typical of $\nu(\text{-}COO^-)$ symmetric and anti-symmetric stretching, respectively. The band at 1312 cm^{-1} is due to ν(C-N) stretching. The bands at 1266 and 1115 cm^{-1} are attributed to $\delta(NH_2)$ symmetric and asymmetric deformations. The bands at 776 and 676 cm^{-1} are ascribed to $\delta(COO^-)$ out of the plane symmetric and asymmetric deformation. Concerning the polymer, the FT-IR spectrum of pure poly-3-aminobenzoate undergoes significant changes, especially in the area relative to the vibrations of the aromatic ring and the functional group NH_2. The bands due to $(\text{-}COO^-)$ in the plane stretching at the 1700 and 1400 cm^{-1}, and (NH_2) out of the plane deformation at 1698, 1634, 1566, 1509 and 1441 cm^{-1} are fairly broad. The broad overlapping bands in the region between 1300 and 1110 cm^{-1}, refer to δ(NH_2) symmetric and asymmetric. Alterations in the spectrum of the polymer are expected because the amine and *p*-methylenes groups of the 3-aminobenzoate molecules interact during polymerization. As for the heat-treated nanocomposites, there is virtually no changes in the profile of the spectra. The bands in the regions between 1700 and 1360 cm^{-1}, related to stretching of the carboxylate group and the aromatic ring are displaced and broader. The bands relative to ν(-COO^-) symmetric and anti-symmetric stretching can be observed in the regions near 1554 and 1384 cm^{-1}. In the region between 1300 and 1110 cm^{-1}there is a shoulder around 1303 cm^{-1} and weak at 1266 cm^{-1}, corresponding to (NH_2) symmetric and asymmetric deformation. Analysis of the bands in the regions below 1200 cm^{-1} is highly compromised because of the large overlap of bands with medium and weak intensity. The bands in the regions below 700 cm^{-1} are due to metal-oxygen-metal vibrations occurring in the inorganic host matrix.

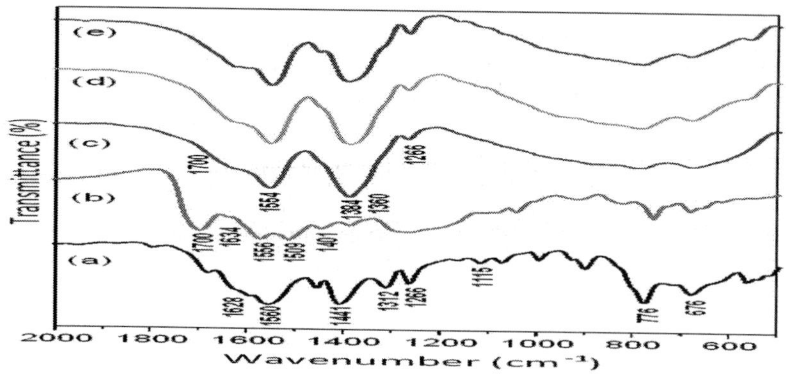

Figure 9: FTIR spectra of (a) sodium 3-aminobenzoate; (b) pure sodium poly-3-aminobenzoate; (c) MgAl-3-aminobenzoate-LDH; (d) MgAl-3-aminobenzoate-LDH with hydrothermal treatment; and (e) MgAl-3-aminobenzoate-LDH heat-treated at 160 °C [61].

Ultraviolet/Visible (UV-VIS) Spectroscopy

The UV-Vis spectra are collected between 200 and 800 nm. Samples are prepared by dissolution of the material in concentrated HCl and subsequent dilution in water.

Figure 10 depicts the UV-Vis spectra pure sodium Poly-3-aminobenzoate and ZnAl-AMB-LDH with different Zn:Al molar ratios(2:1, 3:1 and 6:1) [60].

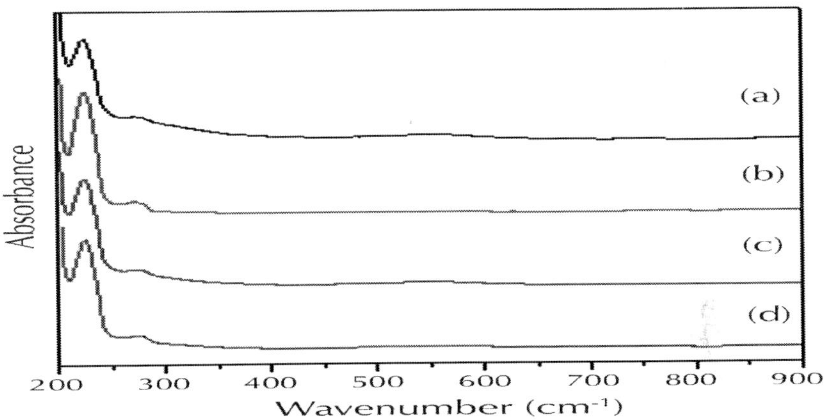

Figure 10: UV-Vis absorption spectra of the materials prepared with (a) pure sodium poly-3-aminobenzoate; (b) Zn_2Al-3-aminobenzoate-LDH; (c) Zn_3Al-3-aminobenzoate-LDH; and (d) Zn_6Al-3-aminobenzoate-LDH.

All the LDH display a band at about 225 nm, after polymerization, the band verified for the monomers is dislocated to lower wavelengths ~215 nm, and a band at ~ 275 nm appears. The latter band is less pronounced for Zn_6Al-3-AMB-LDH prepared by anion exchange in double phase, which is attributed to the n-ϖ* transition due to the presence of non-shared electrons in the COO^- group. After polymerization a peak at ~ 315 nm ascribed to ϖ-ϖ* transition related to conjugation of rings in the polymeric chain is detected. As for the LDH, the first absorption peak intensifies ongoing from the compounds prepared with Zn/Al ratios of 2:1 and 3:1 to 6:1. In the case of the materials prepared by exchange in double phase only for the compounds with Zn/Al ratios of 2:1 and 3:1 the band intensifies. The compound with Zn/Al ratio of 6:1 has the least intense peak.

Transmission Electron Microscopy (TEM)

The best TEM images are generally achieved when LDHs are dispersed in an epoxy resin, centrifuged, and kept at 70 °C for the 72 h, for drying.

After drying, the materials are cut in an ultra-microtome and transferred to hexagonal copper bars appropriated for TEM image acquisition. An alternative approach is to prepare a suspension containing ethanol and LDH. The copper grid is then immersed into the suspension and dried at ambient temperature.

Figure 11 reveals very orderly particles in which the darkest lines represent the inorganic layers and the clearest lines refer to the intercalated conductive polymers [54]. There is good pillaring of the sheets, with a large sequence of darker lines. The basal spacing value estimated from the TEM images can be compared with the one obtained PXRD analysis.

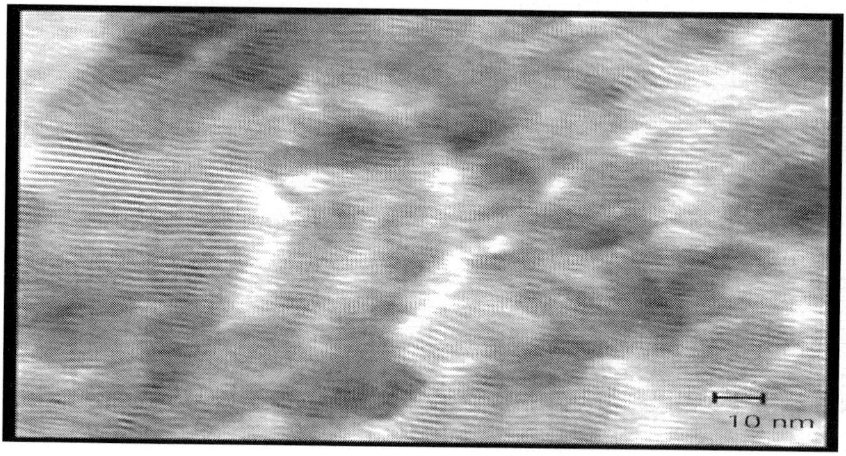

Figure 11: TEM micrographs for ZnAl-4-(1H-pyrrol-1-yl)benzoate-LDH with hydrothermal treatment.

Scanning Electron Microscopy (SEM)

The morphology of the crystallites and nanocomposite particles can also be analyzed by SEM.

For these analyses, the samples are usually supported on the sample port by powder dispersion on double-sided conductive adhesive tape. Because LDHs do not present enough conductivity for generation of good images it is necessary to cover the samples with gold before the measurements, using a sputter equipment.

Figure 12 shows the SEM images of LDH intercalated with 3-aminobenzoate. There is superposition of the sheets, with formation of aggregates on the surface of the cristallyte [60].

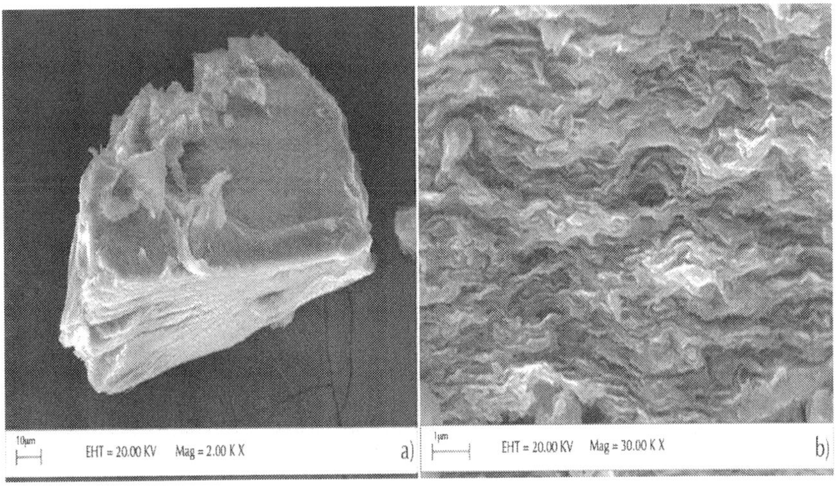

Figure 12: SEM images of ZnAl-3-aminobenzoate-LDH [60].

Cyclic Voltammetry (CV)

Cyclic Voltammetry (CV) experiments are conducted on potentiostats. The supporting electrolyte is 0.1 mol/dm^3 $LiClO_4$ solution, and a conventional electrochemical cell arrangement involving three electrodes is utilized: Platinum wire as the counter electrode, as the reference electrode (Ag/AgCl/KCl(sat)), and glassy carbon, prepared by dip-coating in an aqueous suspension of monomers intercalated into LDH as the working electrode. The potential of the liquid junction is disregarded. CV experiments enable evaluation of the oxidation and reduction processes of the intercalated monomers. A typical voltammogram of ZnAl-LDH intercalated with 3-aminobenzoate anions is presented in Figure 13. The oxidation process involved in the polymerization of 3-aminobenzoate intercalated into LDH can be noticed. Moreover, Zn^{2+} oxidation can be verified.

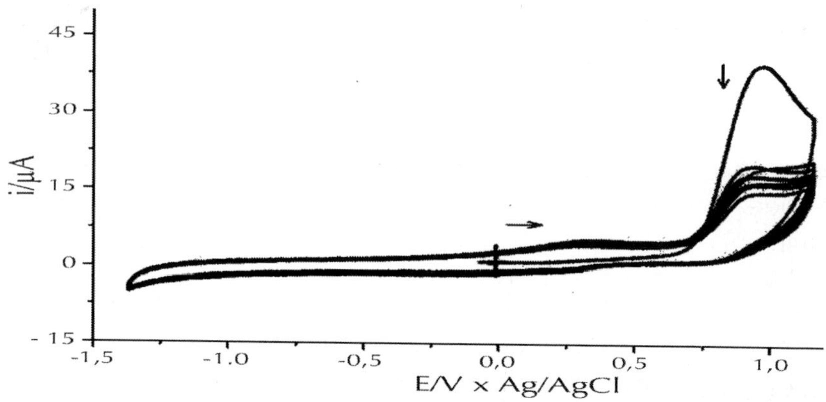

Figure 13: CV for ZnAl-3-aminobenzoate-LDH [60].

There is an irreversible oxidation peak at about 0.960V, and the amplitude of this peak diminishes upon consecutive scanning. This peak is ascribed to 3-aminobenzoate oxidation. A similar behavior has been previously observed for 2-thiophenecarboxylate anions intercalated into ZnAl-LDH.

CONCLUSIONS

Layered Double Hydroxides (LDHs) are materials whose layered architecture enables separation of the inorganic part (double hydroxide), and the organic portion (conductive polymer), thus culminating in a hybrid composite. The *"growth"* of conductive polymers in limited spaces, like the interlayer region of the LDHs, has been shown to be a very promising method for the improvement of the properties of conductive polymers.

On the basis of literature works, it is possible to deduce that, the guests species (monomers) are generally intercalated in a bilayer arrangement within the LDH layers. In this arrangement, the functional groups of the monomers are directed to the inorganic layer, and the aromatic rings occupy the central region of the interlayer spacing. The nature of the substituent group (aliphatic or aromatic) influences the structural organization and the *in situ* polymerization of the resulting hybrid materials.

During the synthesis, some nanocomposites undergo spontaneous polymerization, while others have to be submitted to thermal or electrochemical treatments to reach polymerization. Monomers containing substituents with aliphatic chains, tend to undergo polymerization in milder conditions, because the aliphatic chains provide small mobility of the intercalated monomers, thereby faciliting formation of polyconjugated systems. In some cases, thermal treatment may cause collapse of the layered structure, with consequent formation of oxide.

The thermogravimetric analysis data show that, compared with the pure polymer, the LDH-intercalated conducting polymer is more thermally stable. This stability is provided by the inorganic coverage offered by the LDH layers.

In the case of materials intercalated with conducting polymers, there is initial removal of one electron from the polymeric chain, e.g, through *p* doping. This results in the formation of an electronic state denominated polaron. Generation of the polaron can also be interpreted as ϖ electron redistribution. Moreover, the formation of this entity is associated with distortion of the polymeric chain, which transforms the aromatic form into the quinoid form. The production of polaron may be also due the presence of electronic state located in the energy region found in the middle of gap. The quinoid structure presents smaller ionization energy and larger electronic affinity than the aromatic form. Polaron is chemically defined as a radical ion of spin = 1/2. As the concentration of polarons increases, they tend to recombine, stabilizing the structure and forming a "bipolaron". "Bipolaron" is defined as a pair of equal diamagnetics dication with spin equal to 0 and equal charges. The formation of "Bipolaron" is associated with strong distortion to the LDH net work.

ACKNOWLEDGEMENTS

This work was supported by the Brazilian agencies: Fundação de Amparo à Pesquisa do Estado de Minas Gerais (FAPEMIG), Fundação de Amparo à Pesquisa do Estado de São Paulo (FAPESP), and Conselho Nacional de Desenvolvimento Científico e Tecnológico (CNPq).

REFERENCES

1. Pinnavaia TJ, Beall GW. Eds. Polymer-Clay Nanocomposites; Jonh Wiley & Sons Ltd.: New York; 2000.
2. Alexandre M, Dubois P. Polymer-Layered Silicate Nanocomposites: Preparation, Properties and Uses of a New Class of Materials. Materials Science & Engineering R-Reports 2000;289(1-2) 1-63.
3. LeBaron PC, Pinnavaia TJ. Clay Nanolayer Reinforcement of a Silicone Elastomer. Chemistry of Materials 2001;13(10) 3760-3765.
4. Lagaly G. Introduction: From Clay Mineral-Polymer Interactions to Clay Mineral-Polymer Nanocomposites. Applied Clay Science 1999;15(1-2) 1-9.
5. Darder M, Aranda P, Ruiz AI, Fernandes FM, Ruiz-Hitzky E. Design and Preparation of Bionanocomposites Based on Layered Solids with Functional and Structural Properties. Materials Science and Technology 2008;24(9) 1100-1110.
6. Podsiadlo P, Kaushik AK, Arruda EM, Waas AM, Shim BS, Xu J, Nandivada H, Pumplin BG, Lahann J, Ramamoorthy A, Kotov NA. Ultrastrong and Stiff Layered Polymer Nanocomposites. Science 2007;318(5847) 80-83.
7. Merkel TC, Freeman BD, Spontak, RJ, He Z, Pinnau I, Meakin P, Hill AJ. Ultrapermeable, Reverse-Selective Nanocomposite Membranes. Science 2002;296(5567) 519-522.
8. Ogara JE, Ding J, Walsh D; Waters investments ltd (WATE-Non-standard) Waters technologies corp (wate-non-standard), assignee. Hybrid inorganic-organic material used for separation devices, comprises polymerized scaffolding nanocomposite containing scaffolding functionality capable of chemically interacting with surface of another material patent WO2004105910-A2; GB2419886-A; DE112004000906-T5; US2007141325-A1; JP2007515503-W; GB2419886-B; WO2004105910-A3; JP2012042477-A WO2004105910-A2 09 Dec 2004 B01D-000/00 200505.
9. Chen H, Yu H, Hsiao W, Chen X, Xiao W, You X; Du Pont Apollo Ltd (Dupo), assignee. Backsheet for a photovoltaic module, comprises a nanocomposite layer comprising a polymeric matrix

including a polymer consisting of polyethylene terephthalate and silicate nanoparticles, polymeric layers, and adhesive layers patent US2011259415-A1; CN102280505-A US2011259415-A1 27 Oct 2011 H01L-031/0216 201172.

10. Unnikrishnan L, Mohanty S, Nayak SK, Singh N. Synthesis and Characterization of Polysulfone/Clay Nanocomposite Membranes for Fuel Cell Application. Journal of Applied Polymer Science 2012;124(SI) E309-E318.
11. Yuan S, Li Y, Zhang Q, Wang H. ZnO/Mg-Al Layered Double Hydroxides as Strongly Adsorptive Photocatalysts. Research on Chemical Intermediates 2009;35(6-7) 685-692.
12. Cavani F, Clause O, Trifiro F, Vaccari A. Anionic Clays with Hydrotalcite-Like Structure as Precursors of Hydrogenation Catalysts. Advances in Catalyst Design, 1991; 186-190.
13. Vial S, Prevot V, Leroux F, Forano C. Immobilization of Urease in ZnAl Layered Double Hydroxides by Soft Chemistry Routes. Microporous and Mesoporous Materials. 2008;107(1-2) 190-201.
14. Hsu SL, Chang K, Shiu L, Jang G; HSU S L (HSUS-Individual) CHANG K (CHAN-Individual) UNIV NAT CHENG KUNG (UNCK), assignee. Fabricating Polybenzoxazole Clay Nanocomposite for Microelectronic Industry, by Ion Exchange Reaction of Modifying Agent and Layered Clay, and Polycondensation Reaction of Diacid Chloride and Bis(o-aminophenol) Monomer Patent US2003139513-A1; TW576855-A; US7081491-B2 US2003139513-A1 24 Jul 2003 B29C-039/14 200373.
15. Bonifacio LS, Gordijo CR, Constantino VRL, Silva DO, Kiyohara PK, Araki K, Toma HE. Optical Changes and Writing on Hydrotalcite Supported Gold Nanoparticles. Journal of Nanoscience and Nanotechnology. 2008;8(1) 274-279.
16. Carretero MI, Pozo M. Clay and Non-Clay Minerals in the Pharmaceutical and Cosmetic Industries Part II. Active ingredients. Applied Clay Science. 2010;47(3-4) 171-181.
17. Cunha VRR, Ferreira AMD, Constantino VRL, Tronto J, Valim JB. Layered Double Hydroxides: Inorganic Nanoparticles for Storage and Release of Species of Biological and Therapeutic Interest. Quimica Nova. 2010;33(1) 159-171.

18. Bernardo E, Colombo P, Hampshire S. Advanced Ceramics from a Preceramic Polymer and Nano-Fillers. Journal of the European Ceramic Society. 2009;29(5) 843-849.
19. Okada A, Fukushima Y, Kawasumi M, Inagaki S, Usuki A, Kurauchi T, Kamigaito O, Sugiyama S. Toyota Chuo Kenkyushok, assignee. Composite with High Strength and Excellent High Temp. Properties has Layers of Silicate Mineral Homogeneously Dispersed in a Poly-Amide Matrix. patent DE3632865-A1; JP62074957-A; JP62252425-A; US4739007-A; JP7309942-A; JP7310012-A; JP96022946-B2; DE3632865-C2; JP2663113-B2; JP2724547-B2; US37385-E.
20. Usuki A, Kojima Y, Kawasumi M, Okada A, Fukushima Y, Kurauchi T, Kamigaito O. Synthesis of Nylon 6-clay Hybrid. Journal of Materials Research. 1993;8(5) 1179-1184.
21. Kojima Y, Usuki A, Kawasumi M, Okada A, Fukushima Y, Kurauchi T, Kamigaito O. Mechanical-Properties of Nylon 6-clay Hybrid. Journal of Materials Research. 1993;8(5) 1185-1189.
22. Yano K, Usuki A, Okada A. Synthesis and Properties of Polyimide-Clay Hybrid Films. Journal of Polymer Science Part A-Polymer Chemistry. 1997;35(11) 2289-2294.
23. Leroux F, Besse JP. Polymer Interleaved Layered Double Hydroxide: A New Emerging Class of Nanocomposites. Chemistry of Materials. 2001;13(10) 3507-3515.
24. Wang G, Cai FL, Si LC, Wang ZQ, Duan X. An Approach Towards Nano-Size Crystals of Poly(acrylic acid): Polymerization Using Layered Double Hydroxides as Template. Chemistry Letters. 2005;34(1) 94-95.
25. Wang GA, Wang CC, Chen CY. The Disorderly Exfoliated LDHs/PMMA Nanocomposite Synthesized by In Situ Bulk Polymerization. Polymer. 2005;46(14) 5065-5074.
26. Darder M, Lopez-Blanco M, Aranda P, Leroux F, Ruiz-Hitzky E. Bio-nanocomposites based on layered double hydroxides. Chemistry of Materials. 2005;17(8) 1969-1977.
27. Cavani F, Trifiro F, Vaccari A. Hydrotalcite-Type Anionic Clays: Preparation, Properties and Applications. Catalysis Today. 1991;11(2) 173-301.

28. Roussel H, Briois V, Elkaim E, de Roy A, Besse JP. Cationic Order and Structure of [Zn-Cr-Cl] and [Cu-Cr-Cl] Layered Double Hydroxides: A XRD and EXAPS Study. Journal of Physical Chemistry B. 2000;104(25) 5915-5923.
29. Fogg AM, Dunn JS, O'Hare D. Formation of Second-Stage Intermediates in Anion-Exchange Intercalation Reactions of the Layered Double Hydroxide [LiAl2(OH)(6)]Cl Center Dot H2O as Observed by Time-Resolved, In Situ X-ray Diffraction. Chemistry of Materials. 1998;10(1) 356-360.
30. Pisson J, Taviot-Gueho C, Israeli Y, Leroux F, Munsch P, Itie JP, Briois V, Morel-Desrosiers N, Besse, JP. Staging of Organic and Inorganic Anions in Layered Double Hydroxides. Journal of Physical Chemistry B. 2003;107(35) 9243-9248.
31. Kaneyoshi M, Jones W. Exchange of Interlayer Terephthalate Anions from a Mg-Al Layered Double Hydroxide: Formation of Intermediate Interstratified Phases. Chemical Physics Letters. 1998;296(1-2) 183-187.
32. Morpurgo S, LoJacono M, Porta P. Copper-Zinc-Cobalt-Aluminium-Chromium Hydroxycarbonates and Mixed Oxides. Journal of Solid State Chemistry.1996;122(2) 324-332.
33. de Roy A, Forano C, El Malki K, Besse JP. In Synthesis of Microporous Materials; Ocelli, M. L.; Robson, M. E. (ed.) Van Nostrand Reinhold: New York; 1992. p. 108-169.
34. Lopez T, Bosch P, Ramos E, Gomez R, Novaro O, Acosta D, Figueras F. Synthesis and Characterization of Sol-Gel Hydrotalcites. Structure and Texture. Langmuir. 1996;12(1) 189-192.
35. Reichle WT. Synthesis of Anionic Clay-Minerals (Mixed Metal-Hydroxides, Hydrotalcite). Solid State Ionics. 1986;22(1) 135-141.
36. Taylor RM. The Rapid Formation of Crystalline Double Hydroxy Salts and Other Compounds by Controlled Hydrolysis. Clay Minerals. 1984;19(4) 591-603.
37. Indira L, Dixit M, Kamata PV. Electrosynthesis of Layered Double Hydroxides of Nickel with Trivalent Cations. Journal of Power Sources. 1994;52(1) 93-97.
38. Kooli F, Depege C, Ennaqadi A, de Roy A, Besse JP. Rehydration of Zn-Al Layered Double Hydroxides. Clays and Clay Minerals. 1997;45(1) 92-98.

39. Crepaldi EL, Pavan PC, Valim JB. A New Method of Intercalation by Anion Exchange in Layered Double Hydroxides. Chemical Communications. 1999;2155-156.
40. Kumar D, Sharma RC. Advances in Conductive Polymers. European Polymer Journal. 1998;34(8) 1053-1060.
41. Syed AA, Dinesan MK. Polyaniline – A Novel Polymeric Material - Review. Talanta. 1991;38(8) 815-837.
42. Fritsche, J. Ueber das Anilin, Ein Neues Zersetzungsproduct des Indigo. Journal für Praktische Chemie, 1840; 20 453-459.
43. Letheby HJ. On the Production of a Blue Substance by the Electrolysis of Sulphate of Aniline. Journal of the Chemical Society, 1862;15 161-163.
44. Gospodinova N, Terlemezyan L. Conducting Polymers Prepared by Oxidative Polymerization: Polyaniline. Progress in Polymer Science. 1998;23(8) 1443-1484.
45. Runge F.F. Ueber Einige Produkte der Steinkohlendestillation. Annalen der Physik. 1834; 31 65-78.
46. Skotheim TA, editor. Handbook of Conducting Polymers. M. Dekker: New York, 1998.
47. Meyer V. Ueber Benzole Verschiedenen Ursprungs. Berichte der DeutschenChemischenGesellschaft.1882;15 2893-2894.
48. Challier T, Slade RCT. Nanocomposite Materials - Polyaniline-Intercalated Layered Double Hydroxides. Journal of Materials Chemistry. 1994;4(3) 367-371.
49. Isupov VP, Chupakhina LE, Ozerova MA, Kostrovsky VG, Poluboyarov VA. Polymerization of m-NH2C6H4COO Anions in the Intercalation Compounds of Aluminium Hydroxide [LiAl2(OH)(6)][m-NH2C6H4COO] center dot nH(2)O. Solid State Ionics. 2001;141(SI) 231-236.
50. Moujahid EM, Dubois M, Besse JP, Leroux F. Role of Atmospheric Oxygen for the Polymerization of Interleaved Aniline Sulfonic Acid in LDH. Chemistry of Materials. 2002;14(9) 3799-3807.
51. Moujahid EM, Leroux F, Dubois M, Besse JP. In Situ Polymerization of Monomers in Layered Double Hydroxides. Comptes Rendus Chimie. 2003;6(2) 259-264.
52. Tronto J, Sanchez KC, Crepaldi EL, Naal Z, Klein SI, Valim JB. Synthesis, Characterization and Electrochemical Study of Layered

Double Hydroxides Intercalated with 2-Thiophenecarboxylate Anions. Journal of Physics and Chemistry of Solids. 2004;65(2-3) 493-498.

53. Moujahid EM, Dubois M, Besse JP, Leroux F. In situ Polymerization of Aniline Sulfonic Acid Derivatives into LDH Interlamellar Space Probed by ESR and Electrochemical Studies. Chemistry of Materials. 2005;17(2) 373-382.

54. Tronto J, Leroux F, Crepaldi EL, Naal Z, Klein SI, Valim JB. New Layered Double Hydroxides Intercalated with Substituted Pyrroles. 1. In Situ Polymerization of 4-(1H-pyrrol-1-yl)benzoate. Journal of Physics and Chemistry of Solids. 2006;67(5-6) 968-972.

55. Tronto J, Leroux F, Dubois M, Taviot-Gueho C, Naal Z, Klein SI, Valim, JB. New Layered Double Hydroxides Intercalated with Substituted Pyrroles. 2. 3-(Pyrrol-1-yl)-Propanoate and 7-(pyrrol-1-yl)-Heptanoate LDHs. Journal of Physics and Chemistry of Solids. 2006;67(5-6) 973-977.

56. Tronto J, Leroux F, Dubois M, Borin, JF, Graeff, CFD, Valim JB, Hyperfine Interaction in Zn-Al Layered Double Hydroxides Intercalated with Conducting Polymers. Journal of Physics and Chemistry of Solids. 2008;69(5-6) 1079-1083.

57. Tian X, Wei M, Evans DG, Rao G, Yang H. Controlled Polymerization of Metanilic Anion within the Interlayer of NiAl Layered Double Hydroxide. Clays and Clay Minerals. 2006;54(4) 418-425.

58. Tian X, Wei M, Evans DG, Rao G, Duan X. Tentative mechanisms for In Situ Polymerization of Metanilic Acid Intercalated in MgAl Layered Double Hydroxide Under Nitrogen Atmosphere. Advanced Materials Research.2006;11-12295-298.

59. Wei M, Tian X, He J, et al. Study of the In Situ Postintercalative Polymerization of Metanilic Anions Intercalated in NiAl-Layered Double Hydroxides under a Nitrogen Atmosphere. European Journal of Inorganic Chemistry. 2006;17 3442-3450.

60. Bordonal AC. Materiais Híbridos Orgânico-Inorgânicos: Polímeros Condutores Intercalados em Compostos Lamelares. MS thesis. Universidade de São Paulo; 2012.

61. Tronto J. Síntese, Caracterização e Estudo das Propriedades de Hidróxidos Duplos Lamelares Intercalados com Polímeros Condutores. PhD thesis. Universidade de São Paulo; 2006.

Chapter 6

Physicochemical Analysis and Synthesis of Nonstoichiometric Solids

V. P. Zlomanov[1], A.M. Khoviv[2], and A.Ju. Zavrazhnov[2]

[1]Department of Chemistry, Moscow State University, Moscow, Russia
[2]Department of Chemistry, Voronezh State University, Voronezh, Russia

INTRODUCTION

Technological progress is intimately associated with creation of new materials, such as composites, piezoelectrics, ferroelectrics, semiconductors, superconductors and nanomaterials with preset functional properties. For the synthesis of these materials, it is necessary to study the chemical processes that lead to the composition, structure, and accordingly desired properties. Investigation of the interrelation between the composition, structure and properties of matter and determination of synthetic conditions for obtaining substances with preset composition and structure are the major problems of physicochemical analysis. Although significant progress has been made in understanding several challenges remain for further advancements. These challenges and new approaches include some definitions – stoichiometry, nonstoichiometry, deviation from stoichiometry, substance, phase, component as well as the use of phase diagrams in selecting conditions for the synthesis of nonstoichiometric solids. Since nonstoichiometry is associated with defects, attention is also paid to their classification and formation. Synthesys of solid involves control over phase transformations. For this reason some features of the *P–T–x* phase diagrams are discussed. The following inportant features of *P–T–x* phase diagrams will be considered: highest maximal melting point $T^{max}_{m,AB}$ Of the solid compound S_{AB} , noncoincidence of the solid, liquid and vapor compositions ($x^L \neq x^S \neq x^V$) at this temperature, factors determining the value of the homogeneity range, some features of the terms congruent and incongruent phases and phase processes. Criteria for evaluating the homogeneity of nonstoichiometric solids are also considered.

The term *Physicochemical analysis* was introduced in [1] and defined as the field of chemistry dealing with the interrelation between the composition and properties of matter [1]. The foundations of physicochemical analysis were in [1-6].

It has been realized that it is necessary to study the thermodynamic properties of solids and the phase diagrams of the systems in which these solids occur. Because the properties of solids depend significantly on their composition, great attention has been focused on the physicochemical analysis and foundations of the directed synthesis.

THE ESSENCE OF PHYSICOCHEMICAL ANALYSIS — SOME DEFINITIONS

Investigation of the interrelation between the composition, structure, and properties as well as determination of synthetic conditions for obtaining solids with preset composition and structure are the the basic problems of physicochemical analysis.

The subject of chemistry is the conversions of substances. What is a substance, and what is its conversion? A substance is a multitude of interacting particles possessing certain characteristics: composition, particle size, structure, and the nature of chemical bonding. It is these characteristics that determine the properties of the substance (Figure 1).

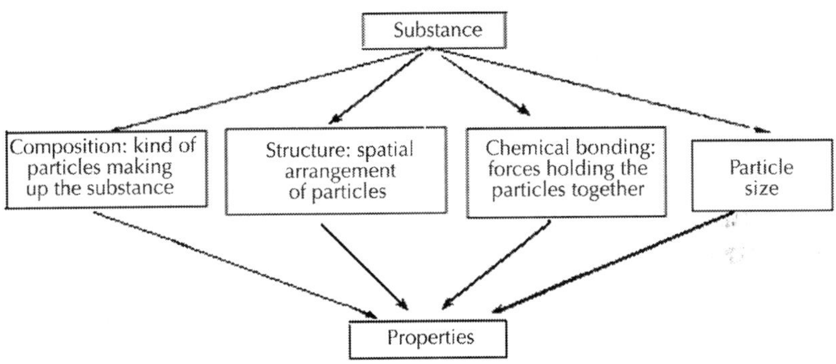

Figure 1: Interrelation between the basic characteristics and properties of a substance.

Composition is defined as the kinds of particles constituting the substance. For example, a sodium chloride crystal is built of sodium and chloride ions occupying cationic sites (Na_{Na}^{+}) and anionic sites (Cl_{Na}^{+}), respectively. The constituent particles can be not only atoms or ions, but also molecules (e.g., I_2 molecules in iodine crystals and water molecules in ice), coordination polyhedra (e.g., SO_4^{2-} tetrahedra in potassium sulfate), and other entities.

Structure is some ordered arrangement of the above particles in space.

The properties of a crystal, such as the lattice energy and electrical, optical, and chemical properties, are determined by the composition and structure of the crystal. Different spatial arrangements of the same particles, such as carbon atoms in diamond and graphite, are characterized by different lattice energies and, accordingly, different properties, including melting and boiling points and hardness.

The chemical bond is understood as the forces binding the particles together. These forces arise from the the Coulomb interaction of electrons and nuclei. Depending on the electron distribution among nuclei, there can be ionic, covalent, and metallic bonds.

The particle interaction energy depends on the particle size. At the nanometer level (1–100 nm), it changes markedly, making possible the formation of new physical and chemical properties of the substance. This is explicable in terms of surface physics and chemistry (dependence of the surface energy on the particle size).

A conversion of a substance is a change in one or several characteristics of the substance (Figure1). This process is accompanied by a change in energy (dU) in the form of heat transfer (δQ) or work execution (δA):

$$dU = \delta Q - \delta A \quad (1)$$

The main purpose of directed synthesis is to obtain substances with the preset composition, structure, and, hence, properties.

The direct synthesis of solids includes control of phase transitions. Therefore, thermodynamic data characterizing phases and their transitions are necessary to estimate the optimum synthetic conditions.

Now let us consider some specific features of the concepts of a phase and a component.

Phase and Component

A substance is made up of particles or their interacting sets with a certain structure and chemical bonding. Energy is an equivalent or measure of these interactions. In thermodynamics, the state of a system is defined using a set of variables, or coordinates, such as pressure P, temperature T, and volume V. For a thermodynamic state of a system characterized by a set of coordinates (intensive thermodynamic properties), Gibbs suggested the term *phase of matter* [7, 8]. This definition emphasizes that the phase of matter is size- and shape-independent. Later, the word combination *phase of matter* gave way to the term *"phase."*

The equation of state of a phase in terms of pressure (P), temperature (T), and composition (x) is written as (2)

$$dG = -SdT + VdP + \sum_{i=1}^{k}\left(\frac{\partial G}{\partial x_i}\right)_{P,T,x_j} dx_i, \quad (2)$$

where G, S, and V are the molar Gibbs energy, molar entropy, and molar volume of the phase, espectively; x_i is the mole fraction of the i-th component k; x_j means the constancy of the mole fractions of all components but not the i-th component. The phases can be individual solid, liquid, gaseous compounds, or solid solutions, including nonstoichiometric compounds, whose thermodynamic properties are described by equations of state like Eq. (2) and are continuous functions of P, T, and $\mathbf{x}$.

Note the following specific features of the concept of a phase. Firstly, existing phases should be distinguished from coexisting phases. The properties (Gibbs energy G, enthalpy, etc.) of an existing phase, such as a solution containing different concentrations of the same substance, are continuous functions of composition. By contrast, the properties of coexisting phases (i.e., phases that are in equilibrium) are equal in all parts of the system. The second specific feature is associated with the definition of a phase as the homogeneous part of a heterogeneous system. This definition is inexact because it does not specify whether homogeneous parts of a heterogeneous system are one phase or different phases. Thirdly, the concept of a phase is broader than the concept of a physical state (gas, solid or liquid). Finally, the fourth specific feature of the concept of a phase is associated with the issue of the minimum set of particles describable in terms of one equation of state like Eq. (2). From the standpoint of kinetic molecular theory, this set should be sufficiently large to allow regular particle energy distribution. Estimates demonstrate that the number of particles must be at least several tens. The properties of such a phase will be affected by surface tension [9].

The components of a system are the types of particles constituting this system. They are called constituents, and their number is designated n. If the concentrations of the n constituents are related by m independent equations, then $k = n - m$ is the number of independent constituents, or, simply, the number of components. For example, the system NH_3(g) + HCl(g) = NH_4Cl(s), which consists of three substances ($n = 3$), whose concentrations are related by one equation ($m = 1$), is two-component since $k = n-m = 3-1 = 2$.

The components are subject to the following constraints:

- their concentrations must be independent of one another;
- they should completely describe the concentration dependence of the properties of the system;
- their number should satisfy the electroneutrality principle.

Stoichiometry, Nonstoichiometry, Deviation from Stoichiometry

The properties of a substance depend on its composition. The great focus of materials sience are the concepts of stoichiometry, nonstoichiometry, and deviation from stoichiometry.

The proportions in which substances react are governed by stoichiometric laws (stoichiometry). These laws, which characterize the composition of chemical compounds, were discovered by systematizing experimental data. The fundamental laws of stoichiometry include the law of constant composition and the law of multiple proportions.

The law of constant composition, established in the 19th century by the French chemist Joseph Louis Proust, states that the chemical composition of a compound is independent of the way in which this compound was obtained. The law of multiple proportions, formulated by the English chemist John Dalton in 1807, states that, when two elements combine with each other to form more than one compound, the mass fractions of the elements in these compounds are in a ratio of prime numbers. Both laws follow from atomistic theory and suggest that the saturation of the chemical bonds is necessary for the formation of a molecule from atoms. Any change in the number of atoms or their nature or arrangement indeed means the formation of a new molecule with new properties.

Are the law of constant composition and the law of multiple proportions always obeyed? They are valid only for the substances constituted by molecules. In fact, the composition of a substance can vary significantly, depending on the preparation conditions. It was long believed that only those chemical substances exist whose composition obeys the law of multiple proportions. They are stoichiometric and are called daltonides in honor of John Dalton. However, as methods of investigation were making progress, it turned out that the properties of solid inorganic substances, such as vapor pressure, electric conductivity, and diffusion coefficients, are composition dependence. In some composition range, the structure, i.e. the arrangement of the components in space remains invariable, while the component concentrations vary continuously. This range is called the homogeneity range or the nonstoichiometry range. These substances are referred to as nonstoichiometric or variable-composition compounds. Earlier, they were called berthollides in honor of Claude Louis Berthollet, Proust's compatriot. A nonstoichiometric compound can be treated as a solid solution of its components, such as cadmium and tellurium in the compound CdTe.

The homogeneity range is characterized by the corresponding deviation from stoichiometry. The stoichiometric composition of a solid compound, e.g., A_nB_m, where n and m are prime numbers, is the composition that obeys the law of multiple proportions. The deviation from stoichiometry or, briefly, nonstoichiometry (Δ) is defined as the difference between the ratio of the number of nonmetal atoms B to the number of metal atoms A in the real $A_nB_{m+\delta}$ crystal ($\delta \neq 0$) and the same ratio in the stoichiometric crystal A_nB_m [10]:

$$\Delta = \frac{m+\delta}{n} - \frac{m}{n} = \frac{\delta}{n}. \tag{3}$$

For the three-component system A–B–C, the composition of the solid phase $(A_{1-x}B_x)_{1-y}C_y$ is conveniently expressed in terms of the mole fraction of the binary compound (x) and nonstoichiometry (Δ).

In this case, the nonstoichiometry Δ can be viewed as the difference between the ratio of the equivalent numbers of nonmetal and metal atoms in the real crystal and the same ratio in the stoichiometric crystal. For example, for $(Pb_{1-x}Ge_x)_{1-y}Te_y$ crystals with a NaCl structure,

$$\Delta = y/(1-y) - 1/1 = (2y-1)/(1-y). \tag{4}$$

The mole fraction (molarity) of the binary compound determines the fundamental properties of nonstoichiometric crystals, including the band gap and heat capacity. The concentration of carriers—electrons and holes—and, accordingly, the galvanomagnetic and optical properties of nonstoichiometric crystals are also associated with nonstoichiometry.

DIRECTED SYNTHESIS OF NONSTOICHIOMETRIC SOLIDS

The strategy of directed synthesis of substances with the preset nonstoichiometry, structure, and properties is based on physicochemical analysis and includes the steps presented in Figure 2.

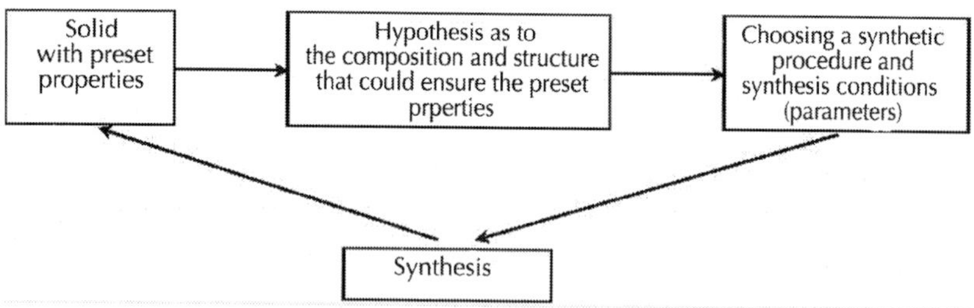

Figure 2: Steps of the directed synthesis of nonstoichiometric solids.

Synthesis is the consequence of processes involved in the conversion of the starting compounds into products. It includes selection and preparation of the starting chemicals (precursors), homogenization of the growth medium (melt, vapor, etc,), the nucleation of the desired phase, nucleus development (growth), and cooling (heating) of the product from the synthesis temperature to room temperature.

Each synthesis step depends on certain conditions or operating parameters. These include the chemical nature and composition of the growth medium, temperature, pressure, diffusion coefficients, heat and mass transfer, and the driving forces of chemical reactions (such as concentration, temperature, pressure, and chemical potential gradients). Figure 3 presents the most important operating parameters of the synthesis.

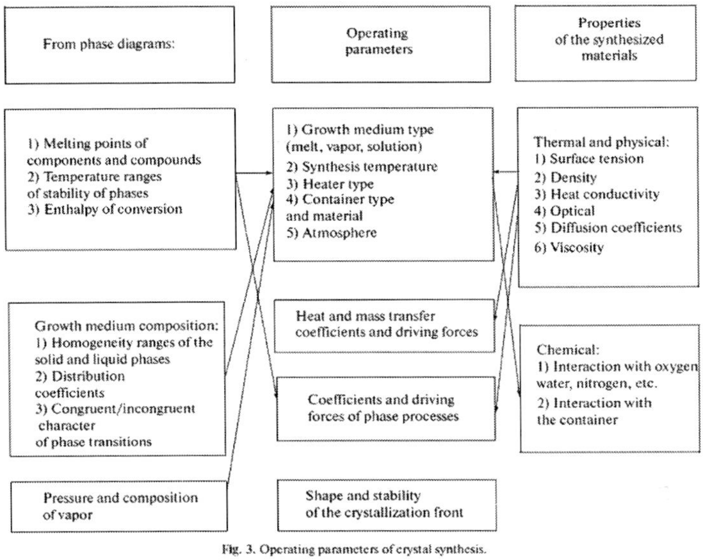

Figure 3: Operating parameters of solid synthesis

Provided that the analytical relationship between the rate of solid synthesis (growth) and the operating parameters is known, this process can be carried out under computer control. This problem has been solved for the Czochralski growth of silicon and germanium crystals and for the Bridgman growth of transition metal oxide crystals [10-14].

Solid synthesis involves phase transitions. For this reason, synthesis conditions are selected using phase diagrams (see section 4). The phase diagram of a system indicates the number of compounds forming in this system and the regions of their stability, specifically, the temperature, pressure, and composition intervals. Thus, using the phase diagram, it is possible to choose the medium (melt, vapor, etc.) for the synthesis of the desired solid, the synthesis conditions (temperature, pressure, and growth medium composition), and the way of carrying out the necessary conversions.

The thermophysical and chemical properties of the starting substances, final synthesis products, and the growth medium to be used (melt, vapor, other solid phases) should be known along with thermodynamic data.

The conversion of a synthesis medium into a solid phase (crystallization) includes the nucleation and development of crystallization centers. Accordingly, it is related with heat and mass transfer and interfacial reaction kinetics. The general problem of analytically describing the crystallization process has not been solved. Furthermore, it is not always possible to evaluate the skill and equipment factors or to establish an unambiguous correlation between the properties of a nonstoichiometric phase and its molecular composition. In this sense, directed synthesis is regarded as being an art [13, 14].

PHASE DIAGRAMS AS A KEY TO SELECTING CONDITIONS FOR SYNTHESIS OF SOLID WITH WELL-DEFINED NONSTOICHIOMETRY

The synthetic conditions for nonstoichiometric solids (S_{AB}) coexisting with a vapor (V) and a melt (L) can be estimated using *P–T–x* diagrams or their *T–x* projections [15-17]. Let us consider the following features of the *T–x* diagrams: the highest (maximal) melting point $T_{m,AB}^{max}$ of the solid S_{AB}; the compositions of the phases at this point; some features of the solidus, liquidus, and vapor lines; the nonvariant points of congruent melting (T_m^c), sublimation (T_S^c), and evaporation (T_e^c); and, finally, congruent and incongruent phases and phase processes.

Maximal Melting Point $T_{m,AB}^{max}$ of a Nonstoichiomertic Solid S

Ab

The Gibbs energy of a phase in a two-component system A–B is given by Eq. (2), so the phase equilibria in this case are represented graphically in a four-dimensional space. This space is explored as four three-dimensional projections: *G–P–T*, *G–P–x*, *P–T–x*, and *G–T–x*.

To clarify the features of the *T–x* diagram let us consider the derivation of part of the *T–x* projection from the *G–T–x* diagram. For this purpose, it should be considered the relative positions of the

G-surfaces of the solid phase S, liquid phase L, and vapor V. In order to determine the arrangement of these three surfaces, we will traverse them with isothermal planes. The projections of the sections of the *G*-surface on the *G*–*x* plane will appear as G^S, G^L, and G^V curves representing the dependence (5) of the Gibbs energy (per gram-atom) on composition at a constant temperature (Figure 4):

$$\left.\begin{aligned} G^S_{A_{1-x}B_x} &= (1-x^S_B)\mu^S_A + x^S_B\mu^S_B \\ G^L_{A_{1-x}B_x} &= (1-x^L_B)\mu^L_A + x^L_B\mu^L_B \\ G^V_{A_{1-x}B_x} &= (1-x^V_B)\mu^V_A + x^V_B\mu^V_B \end{aligned}\right\} \tag{5}$$

where μ^S_A and μ^S_B are the chemical potentials of the components A and B in the solid phase, μ^L_A and μ^L_B are those of A and B in the liquid phase, μ^V_A and μ^V_B are those of A and B in the vapor phase, and x_B is the mole fraction of the component B.

Let us consider the effect of temperature variation on the relative positions of the *G*–*x* curves. At $T_1 > T^{max}_{m,AB}$ (Figure 4a), the solid phase *S AB* is metastable and the two phase system *L* + *V* is stable because a common tangent line can be drawn for the corresponding *G* curves, indicating the equality of the chemical potentials of the components in the equilibrium phases.

As the temperature is decreased, the arrangement of the *G* curves changes. Since the temperature dependence of the Gibbs energy is determined by the entropy $\left(\frac{\partial G}{\partial T}\right) = -S$ and the entropies of the vapor and melt are higher than the entropy of the solid phase, $S^V > S^L > S^S$, the *G* curves shift upwards upon cooling and the G^V and G^L curves do so more rapidly than the G^S curve. As a consequence, a common tangent for the G^S, G^L, and G^V curves can appear at some temperature $T_2 = T^{max}_{m,AB}$ (Figure 4b). This temperature $T = T^{max}_{m,AB}$ is referred as the highest maximal melting point of the solid S_{AB}.

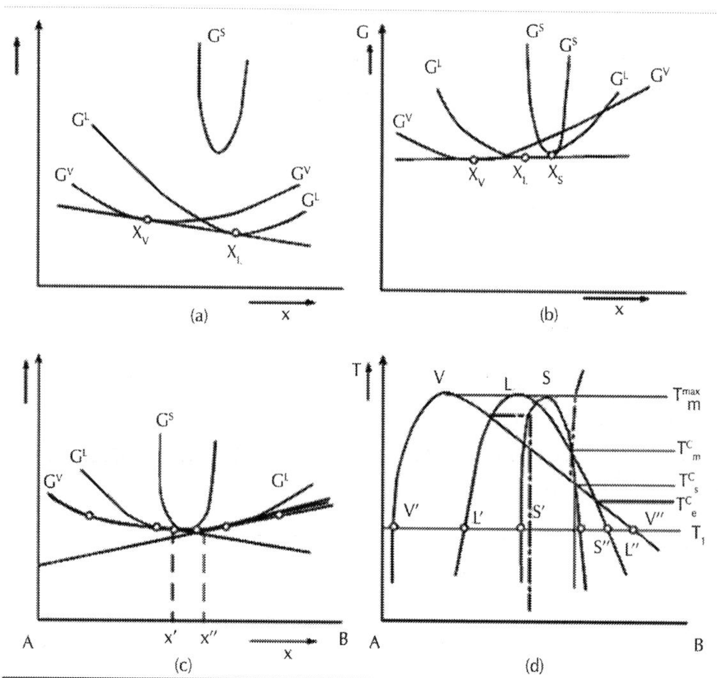

Figure 4: (*G–x*) sections of the *G–T–x* diagram: (a) $T_1 > T^{max}_{m,AB}$ (b) $T_2 = T^{max}_{m,AB}$; (c) $T_3 < T^{max}_{m,AB}$;(d) part of the *T–x* projection near $T^{max}_{m,AB}$.

From the definition of a phase as the totality of the parts of the system whose properties are described by the same equation of state, it follows that the properties of the system are homogeneous functions of composition, pressure, and temperature. Therefore, upon further cooling, for example, to $T_3 < T^{max}_{m,AB}$, it will be possible to draw two tangent lines for each of the G^S_{AB}, G^L, and G^V curves (Figure 4c). The coordinates of the equilibrium phases are designated as *V*, *L*, and *S* in Figure 4d. The subscripts (′) and (″) are given to the phase compositions to the left and right, respectively, of the composition corresponding to $T < T^{max}_{m,AB}$.

As the temperature changes, the G^V-, G^L- and G^S-curves, shifting upwards and to the sides, describe three surfaces: $G^V = G^V(T, x)$, $G^L = G^L(T, x)$ and $G^S = G^S(T, x)$ − and the projections of the tangency points on the *T–x* plane draw the solidus line S›–S››, the liquidus line *L′–L″*, and the vapor line *V′–V″* (Figure 4d). Thus, Figure 4d shows part of the *T–x* projection of the *P–T–x* diagram near the highest melting point of the S_{AB} compound. The points *V′*, *L′*, and S′ and the points *V″*, *L″*, and *S″*, as examples, represent the compositions of the vapor *V*, solid phase S, and liquid phase *L* that are in equilibrium at the temperature $T = T_1$ and are called conjugate points. The lines *VV″*, *LL″*, and*SS″* and the lines *VV′*, *LL′*, *SS′*, as examples, formed by series of these points, are called conjugate lines. The lines connecting conjugate points, such as *V′L′S′* and *V″L″S″*, are called tie lines. Note that, among the various intersection points between the vapor, liquidus, and solidus lines (Figure 4d), only the intersections points of the conjugate lines (*VV″*, *LL″*, *SS″*, *V V ′*, *LL ′*, *SS ′*) have a physical meaning, as distinct from those of the lines *LL ′* и *VV″*, etc.

Specific Features of the Solids, Liquidus and Vapor Lines

Here, let us consider the continuity of the solidus, liquidus, and vapor lines; the factors determining the homogeneity range; the issue of whether this range must include the stoichiometric composition; the causes of the retrograde character of the solidus line; and the concepts of a pseudocomponent and psevdobinary section in multicomponent systems.

The *T–x* and *P–T* projections of the *P–T–x* diagram of a two-component system having a compound *AB* [10, 15] are shown in Figure 5. The Gibbs energy of any phase is a homogeneous function of composition, so the solidus line *S′S″*, the liquidus line *L′L″*, and the vapor line *V′V″*, which represent the temperature dependence of the compositions of the equilibrium phases, are continuous lines having no inflection points.

The homogeneity range of a solid compound is bounded by the solidus line. It is determined by the coordinates of the tangency points of the common tangent line for the *G* curves of the equilibrium phases (Figure 4), i.e., by the equality of the chemical potentials of the components. Therefore, in the general case the homogeneity range (Δ in Eq. (3)) is determined by the relative positions of the G^S, G^L and G^V curves (Figure 4) (i.e., the properties of all coexisting phases) and by the shape of the $(G\!-\!x)_{P,T}$ curves. Thus the nonstoichiometry value depends not only the specific features of the nonstoichiometric solid, namely the radius value, electronic configuration and electronegativity of the constituents species, but on the properties of all coexisting phases.

In the case of the compound *AB* formed from solid components *A* and *B*,

$$A_S + B_S = AB_S;\ \Delta_f G^0 \tag{6}$$

the homogeneity range can be estimated using the following relationship [17]:

$$f(\delta' \ / \ k) - f(\delta''/ \ k) = -\Delta_f G^{\hat{\imath}}, \tag{7}$$

where $\Delta_f G^\circ$ is the standard Gibbs energy of reaction (6), *f* is the monotonic function of composition δ, the subscripts ‹ and ‹› refer to the left and right sides of the homogeneity range, and *k* is a temperature dependent parameter. It follows from Eq. (7) that, the more negative the Gibbs energy $\Delta_f G^\circ$, the larger the difference $f(\delta_{›}/k) - f(\delta_{››}/k)$ and, accordingly, the broader the homogeneity range $\delta_{››} - \delta_{›}$. For crystals dominated by covalent bonding, such as those of the $A^{III}B^{V}$ compounds *GaAs*, *InP*, etc., $\delta_{››} - \delta_{›}$ is on the order of thousandths of an atomic percent. For crystal with polar bonds (*CdTe*, *PbSe*, *SnTe*), the homogeneity range is between a few tenths of an atomic percent and several atomic percent.

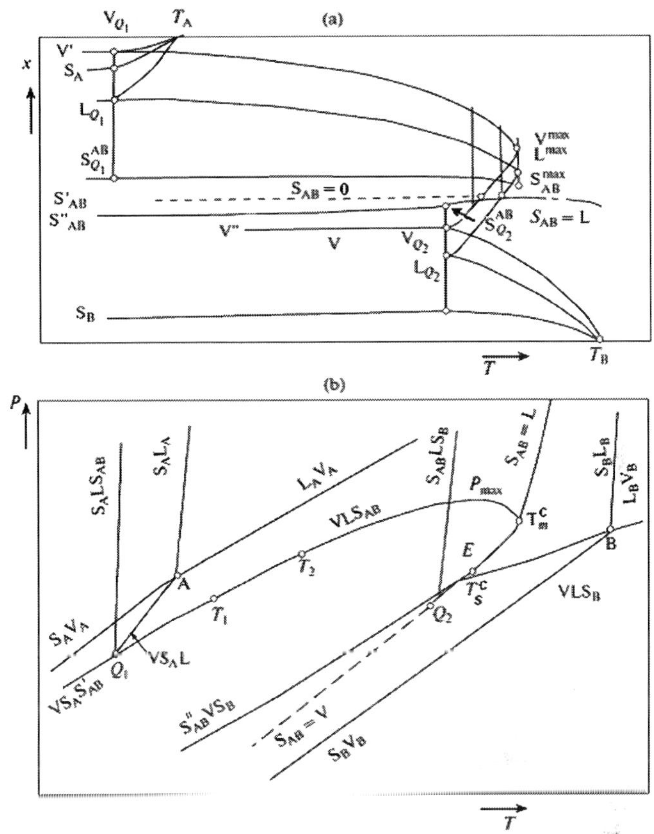

Figure 5: (a) *T–x* and (b) *P–T*- projections of the *P–T–x* diagram of a two-component system with a congruently melting compound (S_{AB}).

The homogeneity range may include the stoichiometric composition or not. Since the $\delta' + \delta''$ value defines the position of the center of the homogeneity range, the expression

$$f(\delta'/k) + f(\delta''/k) = [\mu_B(B) - \mu_A(A)]/RT - [\mu_B(\delta = 0) - \mu_A(\delta = 0)]/RT \qquad (8)$$

describes the dependence of the position of the homogeneity range center on the difference between the Gibbs energies of the pure components (first term in (8)) relative to the isolated atoms in their ground states and on the difference between the chemical potentials of the components in the stoichiometric solid (second difference in (8). If the latter quantity is neglected in a series of crystallochemically similar compounds, the center of the homogeneity range, $(\delta' + \delta'')/2$ will shift in the direction in which the difference $\mu_B(B) - \mu_A(A)$ increases. This can easily be illustrated by examples of nontransition metal chalcogenides.

In the general case, the stoichiometric composition does not correspond to the minimum of the free energy of the solid phase and can fall outside the homogeneity range. When this is the case, the stoichiometric compound does not exist. For example, strictly stoichiometric ferrous oxide can be obtained only at high pressure.

The temperature dependence of the solidus $S'_{AB} S^{max}_{AB} S''_{AB}$ (Figure 5), which defines the boundaries of the homogeneity range of the nonstoichiometric compound S_{AB}, is described by the following equation [16, 18]:

$$\left(\frac{dx^S}{dT}\right)_P = \frac{(\partial S/\partial x)^S_{P,T} - (S^L - S^S)/(x^L - x^S)}{\left(\frac{\partial^2 G}{\partial x^2}\right)_{P,T}} \quad (9)$$

where S^L and S^S are the molar entropies and x^L and x^S are the compositions of the solid phase (S) and melt (L). The solubility of the components in the nonstoichiometric solid $A_{1-x}B_x = A_{1/2-\delta}B_{1/2+\delta}$ can be retrograde: it can initially increase with increasing temperature ($(dx^S/dT) > 0$) and, on passing through a maximum, decrease ($(dx^S/dT)_P < 0$). The maximum solubility $(x^S)_{max}$ is found from the extremum condition for function (9) ($(dx^S/dT)_P = 0$) since $\left(\frac{\partial^2 G}{\partial x^2}\right) > 0$, according to the phase stability criterion. The retrograde solubility is due to the difference between the rates at which the entropies S^L and S^S, including their configurational components, grow with increasing temperature.

The liquidus line $\left(L_{Q_1} L^{max} L_{Q_2}\right)$ and vapor line $\left(V_{Q_1} V^{max} V_{Q_2}\right)$ in Figs. 4 and 5 represent the temperature dependence of the compositions of the melt (L) and vapor (V) in equilibrium with the nonstoichiometric solid S_{AB}.

Noncoincidence of the Phase Compositions at the Maximum Melting Point of a Nonstoichiometric Solid S_{AB}

Both pressure and temperature extrema at T = const and P = const, respectively, exists in a three-phase two-component system provided that the following condition is satisfied [16]:

$$(x^L - x^S)/(x^V - x^S) = (S^L - S^S)/(S^V - S^S) = (V^L - V^S)/(V^V - V^S) \quad (10)$$

Thus, if the coexisting phases have different molar volumes V and entropies S, the composition of one of them will be a linear function of the compositions of the two others: $x^L = \beta x^S + (1 - \beta)x^V$ where β is a coefficient depending on the pressure and temperature and independent of the composition. Since $V^L \neq V^S \neq V^V$ and $S^S \neq S^L \neq S^V$ in the general case, at $T = T^{max}_{m,AB}$ the composition of the solid phase S_{AB} does not coincide with the composition of the melt L and vapor V: $x^S \neq x^L \neq x^V$. This is in agreement with the phase rule:

$$c = k + 2 - r - \alpha, \quad (11)$$

where k is the number of components, 2 is the number of external fields (baric and thermal), r is the number of phases, and α is the number of independent constraints imposed on the intensive variables (c). Indeed, if it is assumed that, at $T = T^{max}_{m,AB}$ $x^S = x^L = x^V$, then α = 2 in Eq. (11) and $c = 2 + 2 - 3 - 2 = -1 < 0$, which is impossible. Unfortunately, this mistake is frequently encountered in the literature [18-21].

In the case of lead, germanium, tin, and cadmium chalcogenides, the vapor phase consists mainly of a chalcogen, so it can be accepted that $x^S - x^V = -0.5$. At $P \cong 1$ atm, $V^S - V^V \cong 10^{-3}\, V^S$ and $V^S - V^L \cong 5{\cdot}10^{-2}\, V^S$. Therefore, $x^S - x^L = 2.5 \times 10^{-5}$; that is, the compositions of the phases at the highest melting point do not coincide, even though they are very similar [14, 22, 23].

Nonvariant Congruent Melting, Sublimation, and Evaporation Points of the Three-phase Equilibrium $S_{AB} + L + V$

Since the solid, liquid, and vapor phases are characterized by different temperature and concentration dependences of the Gibbs energy, the following intersection points of the conjugate liquidus, solidus, and vapor lines can appear on the *T–x* projection. The temperature of the intersection point of the conjugate liquidus and solidus lines refers to solid and liquid phases of equal compositions ($x^S = x^L$) and is called the congruent melting point T_m^c of the S_{AB} phase. The temperature of the intersection point of the solidus and vapor lines refers to a solid phase and vapor of equal compositions ($x^S = x^V$) and is called the congruent sublimation point T_S^c. The temperature of the intersection point of the conjugate liquidus and vapor lines refers to a liquid phase and vapor of equal compositions ($x^S = x^V$) and is called the congruent evaporation point T_e^c. The temperatures T_m^c, T_S^c and T_e^c do not correspond to the stoichiometric composition ($\delta = 0$) of the solid phase $A_{1/2-\delta}B_{1/2+\delta}$, where $x = 1/2 + \delta$ and δ is the deviation from its stoichiometric composition *AB*. For example, congruently melting lead telluride contains (2.8 ± 0.3) 10^{-4} mol excess Te per mole of *PbTe* [22, 23]. Because the formation energies of the defects responsible for the nonstoichiometries $\delta > 0$ and $\delta < 0$ are different, the G^S curve (Figure 4a, 4b, 4c) is asymmetric relative to the $\delta = 0$ composition. Accordingly, the coordinates of the common tangency points of the *G* curves in the general case do not coincide with the $\delta = 0$ composition.

The equality of the compositions of two phases of the three ones involved in the equilibrium means the appearance of one more relationship $\alpha = 1$ between independent variables (degrees of freedom) in the phase rule expression (11). Therefore, the equilibrium at the points T_m^c, T_S^c and T_e^c is nonvariant: $c = 2+2-3-1 = 0$. The solid S_{AB}, which has a homogeneity range at these points, can be considered a pseudocomponent whose properties are composition-independent, and the sections through ternary, quaternary, and other multicomponent systems involving the S_{AB} solid can be considered as pseudobinary.

Congruent and Inconguent Phases and Phase Processes

The concept of congruence is of great significance in the synthesis of nonstoichiometric solids because, in the case of noncoincidence between the synthetics medium (vapor, melt) and solids compositions, there are fluxes of rejected material and the corresponding kinetic instability of the crystallization front.

A phase can be congruent and incongruent in different temperature intervals. A phase obtainable from the phases that are in equilibrium with it by mixing them in appropriate proportions is called a congruent phase. A phase that cannot be obtained from the coexisting phases is called incongruent [24,25]. As an example, let us consider the phase S_{AB} (Figure 5a) in the three-phase equilibrium $S_{AB} + L + V$ at different temperatures. Above the congruent melting point $(T^{c}_{m,\circledast AB} < T < T^{max}_{m,AB})$ the solid phase S_{AB} is incongruent. In the temperature range $T^{c}_{S} < T < T^{c}_{m,\circledast AB}$, it is congruent with respect to the vapor and melt. In designations of three phase equilibria, a congruent phase is written in the middle. Thus, in the former case, the three-phase equilibrium is designated VLS_{AB}; in the latter case, $VS_{AB}L$.

The concepts of congruent and incongruent phases should not be confused with the concepts of congruent and incongruent phase processes.

Phase processes are changes in the state of the system such that the masses of some phases increase owing to the decrease in the masses of others without changes in the intensive parameters (temperature, pressure, phase compositions) [26]. A phase process in which one phase forms or disappears is called congruent. A phase transition in which more than one phase forms or disappears is called incongruent.

The same phase, for example, S_{AB} (Figure 5a), can be involved both in congruent and in incongruent phase processes. At a temperature or pressure corresponding to the three-phase line $Q_1 Q_2$ in Figure 5b, the solid phase S_{AB} melts congruently, yielding a melt and a vapor: $S_{AB} = L + V$. The compositions of the phases involved in this congruent phase process are different and are represented by the solidus, liquidus, and vapor lines in the *T–x* projection (Figure 5a). The melt and solid phase compositions coincide only at the congruent melting point T_m^c. At the temperatures corresponding to the eutectic nonvariant points T_{Q_1} and T_{Q_2} (Figure 5a), the S_{AB} phase is involved in the incongruent phase processes

$$V + S_A = L + S_{AB} \text{ and } S_{AB} + L = V + S_B.$$

Thus, S_{AB} phase can be said to melt congruently only in a certain temperature range.

Now let us consider the usage of the terms *congruently melting compound* and *incongruently melting compound* in the literature. Firstly, this differentiation is not strict, because the same compound (e.g.,S_{AB}) can be involved in congruent and incongruent phase processes, depending on the temperature. Sometimes, a congruently melting compound is understood as a compound melting without dissociation or decomposition. However, a solid–melt phase transition is accompanied by the breaking and relaxation of chemical bonds in the crystal and by long-range disordering. Therefore, the term*melting without decomposition* is not quite correct, and it should be understood as the identity of the overall compositions of the coexisting phases. In the strict sense, the overall compositions of the phases coincide only at the congruent melting point $T^c_{m,@AB}$, which is below the highest melting point

$$T^{max}_{m,\,AB} : T^{c}_{m,@AB} << T^{max}_{m,\,AB}$$

.

An incongruently melting compound is sometimes understood as a solid compound that decomposes into a solid phase S_B and a liquid L upon melting (Figure 6). The compositions of the resulting phases differ from the composition of the parent phase. However, in some temperature range, such as $T_{Q_1} < T < T_{p'}$ the "incongruently melting compound" is involved in the congruent melting process $S_{AB} = L + V$.

An essential feature differentiating congruently and incongruently melting compounds is that the highest melting point of a congruently melting compound is higher than the temperatures of the nearest nonvariant points: $T^{max}_{m,AB} > T_{Q_1}$ and T_{Q_2} (Figure 5a).

The highest melting point of an incongruently melting compound is intermediate between the temperatures of the nearest nonvariant points of the system.

Supersaturation and synthesis of a nonstoichiometric solid at a fixed vapor pressure of the volatile component can be produced by cooling or, conversely, heating the three-phase system. The latter case corresponds to the temperature range $T_1 < T < T_2$ in Figure 5b. For example, cadmium telluride crystals were obtained by heating cadmium enriched melts [10]. Note that the composition of the crystals that were grown using the vapor–liquid–crystal technique always lies in the solidus line, i.e., at the boundary of the homogeneity range [15].

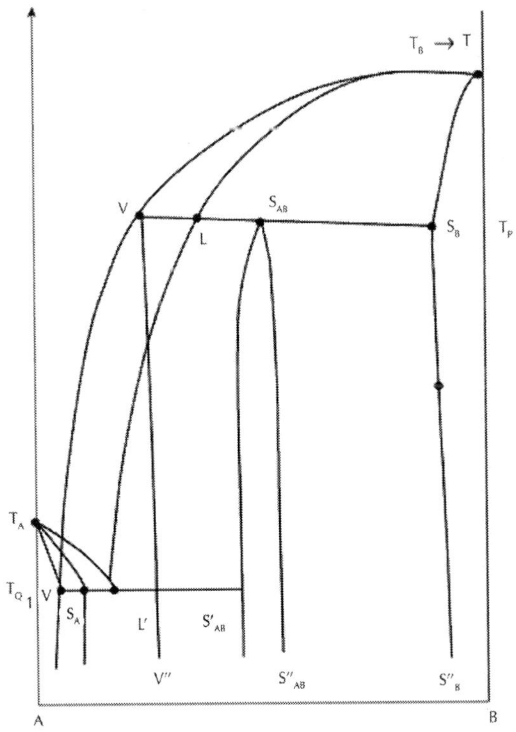

Figure 6: *T–x* projection of the *P–T–x* diagram of a two-component system with an incongruently melting compound (S_{AB}).

NONSTOICHIOMETRY AND DEFECTS IN SOLIDS

Solids like S_{AB} are grown from the vapor, melt (solution), or solid phases, which are called growth media. Crystallization can be viewed as the transfer of atoms of the components *A* and *B* from the growth medium (g.m.) to their regular sites A_A^x and B_B^x in the crystal lattice of the S_{AB} compound:

$$A^{g.m.} = A_A^x + V_B^x;\ \Delta G_1;\ \ K_1 = \frac{[A_A^x][V_B^x]}{a_A} = exp(-\Delta G_1 / kT), \tag{12}$$

$$\hat{A}^{g.m.} = B_B^x + V_A^x\ ;\ \Delta G_2;\ K_2 = \frac{[B_B^x][V_A^x]}{a_B} = exp(-\Delta G_2 / kT) \tag{13}$$

where [ ] designates concentrations and a_A and a_B are the activities of the components in a growth media. The generation of V_B^x and V_A^x vacancies via reactions (12) and (13) is explained by the principle of conservation of the ratio of the numbers of sites characteristic of a given crystal lattice.

Because of the size and energy differences, the Gibbs energies ΔG_1 and ΔG_2, the equilibrium constants of reactions (12) and (13), and, accordingly, the numbers of atoms A and B in the crystals turn out to be different: $K_1 \neq K_2$ and $[A_A^x] \neq [B_B^x]$. Thus, nonstoichiometry appears; that is, the difference between the *B*-to-*A* ratios in the real and stoichiometric crystals becomes nonzero.

Note that the properties of crystals are affected not by the nonstoichiometric atoms A_A^x and B_B^x that occupy their regular sites, but by the defects (resulting from a disordering of the ideal structure).

These defects may be the vacancies V_A^x and V_B^x or the interstitial atoms A_i^x and B_i^x

$$A^{g.m.} = A_i^x + \Delta G_3, \tag{14}$$

$$B^{g.m.} = B_i^x + \Delta G_4. \tag{15}$$

This circumstance is due to the fact that the A_A^x and B_B^x species do not change the energy structure of the crystal, but complete it in a way.

Near the defects (V_A^x, V_B^x, A_i^x, B_i^x), the energy field and, accordingly, the electrical, mechanical and other properties of the crystal are altered (Figure 7). Thus, defects play an important role in the description of the real structure and properties of nonstoichiometric solids.

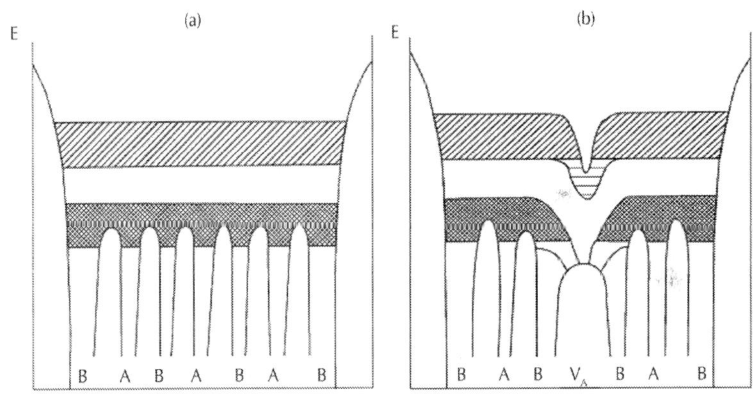

Figure 7: Model of the energy spectrum of (a) an ideal and (b) a nonstoichiometric crystal containing V_A vacancies.

CLASSIFICATION AND FORMATION OF DEFECTS

Defect formation processes and defect classification are significant points in control of the defect composition of solids in their directed synthesis.

An ideal, or perfect, solid is one in which all particles making up the substance or structure elements (atoms, ions, molecules, etc.) occupy their regular sites in the lattice. Under heating, irradiation with a beam of high-energy particles, or mechanical treatment, the regular arrangement of particles over their sites undergoes disordering: some particles can leave their sites. The resulting disorder in the arrangement of particles over their normal sites is called defects [27].

In terms of geometry and size, all defects are divided into point and extended defects [27, 28] (Figure 8).

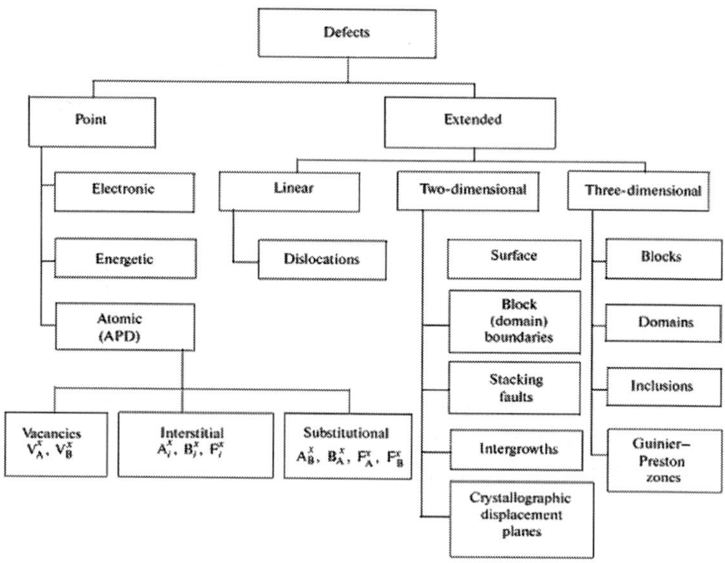

Figure 8: Classification of defects.

The size of point, or zero-dimensional, defects is comparable with the interatomic parameter. The zero-dimensional (OD) defects include electronic defects (holes, electrons, exitons), energy defects (phonons, polarons), and atomic point defects (APDs). The APDs in nonstoichiometric *AB* crystals include V_A^x and V_B^x vacancies (absence of atoms or ions in lattice sites); interstitial atoms A_i^x, B_i^x and F_i^x, where *F* - foreign atoms (the upper symbol *x* means the neutrality of the defect with respect to its environment), the lower symbol means the defect location.

The formation of an APD is an endothermic process requiring a small amount of energy: $0.5 \leq Ef \leq 3$ eV. Therefore, APDs are equilibrium defects and their concentration depends on synthesis conditions, namely, the temperature and the partial pressures of the components. The size of an APD is 0.1–0.5 nm; however, APDs polarize their environment in the crystal structure, causing slight displacements of neighborions, and largely determine the physical and chemical properties (diffusion, electric conductivity, etc.) of the nonstoichiometric crystal.

Vacancies, interstitial and antistructure (A_B^x, B_A^x) defects are classified as intrinsic defects of a crystal. The concentration (c_j) of these defects takes on a thermodynamically equilibrium value at any $T > 0$ K: $c_j \approx \exp(-W_j/kT)$, where W_j is the defect formation energy, $0 < W_j < 3$ eV. The higher the temperature, the higher the APD concentration $c\,j$. As the crystal is cooled to room temperature, part of the APDs can annihilate via various mechanisms. However, many APDs persist even at exceptionally low reactions (12)–(15), but also from the irradiation of the crystal with high-energy (>1 MeV) electrons at very low temperatures.

Oppositely charged APDs can be attracted to one another to yield new APDs as electroneutral association species, such as (V^-_A V^+_B), (V^-_A F^+_A), $(V_A^x)n$. Dipole–dipole interaction leads to the formation of APD clusters, which can serve as nuclei of other phase in the nonstoichiometric crystal.

Extended defects include linear (1D), or one-dimensional, and two (2D)- and three-dimensional (3D) defects [29, 30] (Figure 8). Let us consider some specific features of these defects.

Linear defects, or dislocations, are similar to point and two-dimensional defects in the sense that their size is comparable with the unit cell parameter. In the third dimension, the dislocations are fairly long or even infinite. The simplest kind of dislocation is the edge dislocation.

The most important two-dimensional (2D) defects include solid surface, block (domain) boundaries, stadling faults and crystallographic displacement planes, which are surfaces in which coordination polyhedra of two contacting ideal are rearranged.

An example of periodic intergrowths is provided by the family of tungsten bronzes, A_xWO_3 (A = alkali metal). In these compounds, WO_3

layers are intergrown with hexagonal bronze bands. The formation of intergrown bronzes is likely favored by the synthesis conditions. The boundaries of intergrowths can be periodic (ordered) or nonperiodic (disordered).

Note the following specific features of two-dimensional defects. Firstly, their formation energy is fairly high (>3 eV) and they are kinetically stabilized nonequilibrium defects. The "frozen" state of these defects is responsible for their "memory" for preparation history. Secondly, these defects do not change the stoichiometry of the substance. Thirdly, planar defects result from APD interaction and exert a significant effect on on the reactivity and physical properties of nonstoichiometric solids.

The size of three-dimentional defects (Figure 8) exceeds the lattice constant in all three directions. These defects are, in essence, macroscopic imperfections of the crystal structure and are formed during crystal growth and subsequent processing. Three-dimentional defects include separate blocks; mosaics (totality of a large number of small- and large-angle boundaries); inclusions (microdeposits) resulting from phase transitions, such as the decomposition of a solid solution; magnetic domains (crystal zones with the same orientation of spins or electric dipoles); Guinier–Preston zones (parallel platelike formations as thick as a few unit cells, separated by different distances and having the same composition as the crystal); cavities; and cracks. Three-dimentional defects can be viewed as resulting from defect association and ordering. For example, pores can be considered to result from the association of a large number of vacancies. Bulk defects also include elastic tensile and compressive stresses.

Deviations from stoichiomrtry may be so large that defect interactions become significant leading to direct ordering, clustering, superstructure formation, long-range ordering, and the formation of new nonstoichiometric phases differing in symmetry, energy and other aspects from the parent phase. In such systems defects are intrinsic components of the crystal structure rather than being statistically distributed imperfections. The crystal-chemical and thermodynamic aspects of nonstoichiometric compounds with narrow and broad homogeneity range as well the approaches for controlling the nonstoichiometry and considered in [31].

SUBSTANCE HOMOGENEITY CRITERION IN PHYSICOCHEMICAL ANALYSIS

The Gibbs energy (*G*) of a crystal is a statistical quantity related to the distribution function. The mean*G* value determines the most likely distribution of zero-, one-, two-, and three-dimensional defects. Fluctuations around the mean value are possible, and, in a closed system (solid substance with a constant net composition at a fixed temperature), the configurational fluctuation can manifest itself as a change in composition within a small region as a consequence of a random motion of particles inside some volume element or to its surface. This brings up the question of what substance can be called homogeneous. The degree of heterogeneity of a solid phase is characterized by the statistical distribution of structure elements from which the crystal is built. These structure elements are the atoms of the components of the system in their regular sites, as well as various zero-, one-, two-, and three-dimensional defects in their regular positions (Figure 8).

A quantitative estimate of the degree of heterogeneity can be based on the following three types of distributions:

- distribution of structure elements in some measurable volume,
- distribution of these volumes in the crystal, and
- distribution of measurement data and properties of the solid phase.

Let σ be the confidence interval and c_i be the concentration of structure elements in the *i*th microvolume. If the solid substance can be called homogeneous.

$$\left|c_i - 1/N\sum_{i=1}^{N} c_i\right| \le \sigma \qquad (16)$$

If there is an *i* value at which this inequality is invalid, the solid phase should be considered to be heterogeneous. For practical use of a material, of significance are such deviations of the property in a given object from the weighted average value for the entire system that go beyond the confidence interval. In this sense, heterogeneity can be understood as the totality of property values, measured in all microvolumes, that fall outside the confidence interval. An analysis of generalized heterogeneity criteria in terms of an autocorrelation function was carried out by Nikitina et al. [32]. A thermodynamic analysis of defect ordering and an interpretation of the concentration dependences of physical properties taking into account short- and long- range order parameters were made using a model based on the above stricture elements or cluster components [33].

Some features of homogeneity criteria, physicochemical analysis, thermodynamics in materials science and inorganic crystal engineering are discussed in [34-37].

CONCLUSIONS

Technological progress is intimately associated with creation of new materials, such as composites, piezoelectrics, ferroelectrics, semiconductors, superconductors and nanomaterials with preset functional properties. For the synthesis of these materials, it is necessary to study the chemical processes that lead to the desired properties

Physicochemical analysis is the field of chemistry dealing with these processes, the interrelation between the composition, structure and properties of matter and determination of synthetic conditions for obtaining such substances. Several challenges and new approaches have been discussed. They include the concepts of a substance, phase, component, directed synthesis strategy as well as some definitions. Attention was paid to nonstoichiometry, classification and formation of defects.

Synthesis of nonstoichiometric solid involves control over phase transformations. The P–T–x phase diagram is a key to selecting conditions for synthesis of solid with well-defined nonstoichiometry. For this reason the following features of *P–T–x* phase diagrams were considered: highest melting point of a nonstoichiometric compound ($T^{max}_{m,AB}$), noncoincidence between the solid-, liquid-, and vapor-phase compositions ($x^L \neq x^S \neq x^V$) at this temperature, factors determining the nonstoichiometry range, the terms congruent and incongruent phases and phase processes. Criteria for evaluating the degree of heterogeneity of nonstoichiometric solids were also considered.

REFERENCES

1. Kurnakov N.S. Zadachi Inst. Physikochemical Analysis. Izv. Inst. Phys.Chem. Analysis. 1919; 1(1), 1-7.
2. Kurnakov N.S. The Introduction to Physicochemical Analysis. 4th ed. Moscow: Akad. Nauk SSSR; 1940 [in Russian].
3. Kurnakov N.S. Collection of Selected Works. Leningrad/Moscow: ORTI, Khimteoret; 1938; vol. II, p. 1-202 [in Russian].
4. Kurnakov N.S. Selected Works. Moscow:Akad. Nauk SSSR; 1963; vol. II, 1-361 [in Russian].
5. Anosov V.Ya., Pogodin S.A. The Principles of Physicochemical Analysis. Moscow: Akad.Nauk SSSR; 1947 [in Russian].
6. Tamman G. The Manual on Heterogeneous Equilibria. Leningrad:ONTI, Khimteoret; 1935 [inRussian].
7. Gibbs J.W. Thermodynamics: Statistical Mechanics. Moscow: Nauka; 1982.
8. Voronin G.F. The Fundamentals of Thermodynamics. Moscow:Moskow State University; 1987 [in Russian].
9. Fuks G.I. The Smallest Piece, the Smallest Drop. Khimya and Zhizn'. 1984; 2, 74-76 [in Russian].
10. Kröger F,A. The Chemistry of Imperfect Crystals. Amsterdam: North-Holland Publ. Co.; 1973.
11. Vasiliev Ya., Akhmetshin R., Borodlev Yu. BGO Crystal Growth by a Low Thermal Gradient Czochralski Technique. Nucl. Instr. Meth. Phys. Res. 1996; A379, 533.
12. GolyshevV., Gonic M. et al. HeatTransfer in Growing Be4Ge3O12 Crystals under Weak Convection. J. Cryst. Growth. 2004; 262, 202-214.
13. Rosenberger R.F. Fundam. Cryst. Growth. New-York: Springer-Verlag; 1979.
14. Gottshtain G. The Physical Foundation of Materials Science. Berlin-Heidelberg: Springer-Verlag; 2004.

15. Zlomanov V.P., Novoselova A.V. P–T–x Diagrams of Metal–Chalcogen Systems. Moscow: Nauka; 1987 [in Russian].
16. Storonkin A.V. The Thermodynamcis of Heterogeneous Systems. Leningrad: Leningrad State University; 1967; part 1 [in Russian].
17. Brebrick R.F. Nonstoichiometry in Binary Semiconductor Compounds M0.5–**δ**N0.5+**δ**. Prog. Solid State Chem. 1966; 3, 213-263.
18. Roozeboom W.B. Die heterogenen Gleichgewichte vom Standpunkte der Phasenlehre. Braunschweig: Vieweg; 1904.
19. Ricci J.E. The Phase Rule and Heterogeneous Equilibrium. New York: Dover; 1966.
20. Khaldoyanidi K.A. Phase Diagrams of Heterogeneous Systems. Novosibirsk: INKh RAN; 1991; part 1 [in Russian].
21. Levinskii Yu.V. p–T–x Diagrams of Binary Metal Systems. Moscow: Metallurgiya; 1990; book 1 [in Russian].
22. Peter K., Wenzel A., Rudolph P. The *p–T–x* Projection of the System *Cd–Te*. Cryst. Res. Technol. 1990; 25(10), 1107-1114.
23. Avetisov I. Kh., Mel'kov A.Yu., Zinov'ev A.Yu., Zharikov E.V. Growth of Nonstoichiometric PbTe Crystals by the Vertical Bridgman Method Using the Axial-Vibration Control Technique. Crystallography Reports. 2005; 50, Suppl. 1, 124-129.
24. Meuerhaffer W., Sanders A.Z. Z. für Phys.Chem. IV. Phys. Chem. 1899; 28, 453.
25. W. Meuerhaffer W., Sanders A.Z. Z. für Phys.Chem. II. Phys. Chem. 1905; 53, 513-518.
26. Münster A. Chemische Thermodynamik. Berlin: Academie – Verlag; 1969.
27. Van Bueren H.G. Imperfections in Crystals. Amsterdam: North-Holland; 1960.
28. Fistul' V.I. The Physics and Chemistry of the Solid State. Moscow: Metallurgiya; 1995 [in Russian].
29. West A. Solid State Chemistry and Its Applications. Chichester: Wiley; 1984.
30. Rabenau A., editor. Problems of Nonstoichiometry. Amsterdam: North Holland; 1970.

31. Zlomanov V.P. Crystal Growth of Nonstoichiometric Compounds. Inorganic Materials. 2006, 42, Suppl. 1, 19-48.
32. Nikitina V.G., Orlov A.G., Romanenko V.N. Problem of Nonhomogeneity Distribution of Atoms and Defects in Semiconductor Crystals. In: Growth of Semiconductor Crystals and Films. Novosibirsk: Nauka; 1981; 204 [in Russian].
33. Men' A.N., Bogdanovich M.P., Vorob'ev Yu.P. et al. Composition–Imperfection–Properties of Solid Phases. Cluster Components Method. Moscow: Nauka; 1977 [in Russian].
34. De Hoff Robert T. Thermodynamics in Materials Science. Taylor & Francis. Boca Raton. London. New-York. 2011. P. 531.
35. Brammer L. Developments in Inorganic Crystal Engineering. Chem. Soc. Rev. 2004. V.33. P. 476-489.
36. Shriver & Atkins. Inorganic Chemistry. Oxford University Press. Fourth Ed. 2006. 822 P.
37. Ashby M., Sherclift H., Ceban D. Materials, Engineering, Science, Processing and Design. Elsevier. Amsterdam. Boston. 2011.

Chapter 7

Materials Assessment and Process Characterization for Lead-Free Soldering

Chien-Yi Huang[1], Chen-Liang Ku[2], Hao-Chun Hsieh[2], Ming-Shu Li[2], Chia-Hsien Lee[2], and Cheng-I Chang[2]

[1]Department of Industrial Engineering and Management, National Taipei University of Technology, Taipei, Taiwan, R.O.C

[2]Process Technology Enabling & Materials Characterization Div., Global Operations, Wistron Corporation, Hsinchu, Taiwan, R.O.C

INTRODUCTION

Trends in the packaging of semiconductors are towards miniaturization and high functionality. Increased functional density and reduced I/O (Input/Output) spacing are the market trends. In the assembly process with surface mount technology (SMT), a solder paste consisting of solder particles and flux is deposited onto the PCB's bonding pads through stencil printing. The electronic components are then placed onto the printed circuit board (PCB) by a pick-and-placement machine. During reflow soldering, the solder paste melts and solidifies to form reliable solder joints. With the continuously increasing demands of electronic appliances, energy consumption and the use of hazardous substances, such as heavy metals and chemical ingredients, have become of serious concern. Traditionally used tin-lead solder is known to be harmful to human life and not compatible with an eco-environment. Environmental protection has become an international issue, and on July 1, 2006, the. European Union (EU) implemented a directive restricting hazardous substances (RoHS). One lead-free solder candidate, SAC305 (Sn96.5/Ag3.0/Cu0.5), has been widely used as a replacement for traditional tin-lead solder (Sn63/Pb37). Industry reports indicate that approximately 50–70% of soldering defects are attributed to the solder paste printing process. A sufficient amount of paste and the desired reflow soldering profile are mandatory for reliable interconnections. Solderability analysis is one of the major approaches in characterizing PCB assembly, quality control and reliability.

First, the procedure of solder paste evaluation is illustrated. In the lead-free process, the flux and solder alloy play important roles in the effectiveness of electrical resistance and the strength of the solder joints. Test items include wettability, spread, solder balls, slump, tackiness, viscosity, copper minor, silver chromate, assembly test and reliability tests. This study demonstrates the use of measuring tools combined with statistical methods to investigate the effects of the laser cutting taper angle and speed on the stencil quality for the fine pitch solder paste printing process. The stencil quality is defined by the amount of stainless steel residue after laser cutting and the roundness of the stencil aperture for a circular aperture design. The solder paste inspection data is analyzed to investigate the influences of electropolishing and the stencil area ratio on the transfer ratio during solder paste deposition. The aim of this study is to determine the optimal parameters for stencil fabrication and the corresponding performance in solder paste deposition. The temperature profiles on the critical locations of the PCB during reflow soldering are also important to achieve the desired solder joint quality. This study establishes a procedure to predict the temperature profile based on given information about the PCB design and component loading. Critical factors, such as the conveyor speed and temperature settings of adjacent zones, that can influence the heating process are identified and investigated. A regression model and artificial neural network are constructed to more accurately predict the temperature profile. Results of this study will help improve the efficiency of the temperature setting process, especially in the pilot run stage. Finally, the analytical techniques used for soldering properties, including visual inspection, side-view microscopy, x-ray inspection and dye staining analysis, are also discussed. The crack size and percentage are classified according to the crack area. In this study, the solderability analysis is carried out to determine whether any cracks in the solder joints occur in the CPU and RAM of the ball grid arrays (BGAs) components. It may cause the failure symptom of no display to appear on an ultrabook while the power is on.

STENCIL EVALUATION OF ULTRA-FINE PITCH SOLDER PASTE PRINTING PROCESS

The surface mount technology (SMT) is commonly used in electronics industry to assemble the components onto the PCB. It (SMT) enables the production with relatively low cost. Stencil printing is known to significantly influence the process quality in Surface Mount Technology (SMT). For ultra-fine pitch applications, solder paste printing acts as one of the most critical steps in the SMT assembly process. A review of the literature indicates that solder paste printing is responsible for around 60% of the total process defects. One of the key factors for the performance of solder paste deposition is the stencil fabrication quality. Other important factors include the stencil aperture design, paste printing parameters and stainless steel raw material. The incoming inspection of the stencil typically includes a frame appearance examination and tension force measurement. Inspection of individual apertures under a microscope is not feasible due to the large size of the stencil. Related research also shows that stencil aperture and wall roughness impact the solder paste deposition quality. Laser cutting is widely used due to its advantageous performance with respect to cost and aperture quality. While electroformed stencils are typically considered for the assembly of 01005 passives, evaluations of the laser cut stencil manufacturing process through solder paste printing optimization are lacking. The stencil manufacturing process generally includes three stages, namely, screen stretching, laser cutting and electropolishing/deburring (Fig. 1).

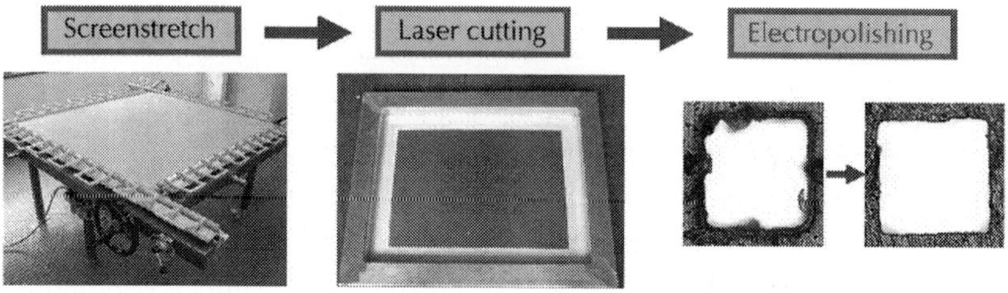

Figure 1: Stencil manufacturing process.

Measuring tools combined with statistical methods were used to investigate the effects of the laser cutting taper angle and speed on the stencil quality. The stencil quality is defined by the amount of stainless steel residue after laser cutting and the roundness of the stencil aperture for a circular aperture design. The solder paste inspection data was analyzed to investigate the influence of electropolishing and stencil area ratio on the transfer ratio during solder paste deposition. Weighted ranking was adopted to indicate the variation in solder paste deposits for 01005 and 0.4/0.3mm pitch CSP pads due to the stencil fabrication process. The aim was to determine the optimal parameters for stencil fabrication and corresponding performance in solder paste deposition.

The laser machine and materials used for stencil aperture cutting and stencil materials were evaluated. Experiments included: (1) material properties assessment – to test the durability of the screen; (2) stencil quality inspection – to evaluate the occurrence of clogging, aperture roundness and taper angle; (3) electropolishing process inspection – to determine the optimal process parameters for laser cutting for various taper angles; and (4) printing process assessment – to evaluate the stencil printing performance for 01005/0201 RLC and 0.4/0.3mm pitch CSP applications.

Experiments

Material Properties Assessment

The durability of various raw materials against IPA cleaning solvent was evaluated. The test samples were 736 mm square and 0.1 mm thick. The stencils (without aperture) were cleaned by the cleaning machine. After cleaning for 90 min, the samples were dried for 10 min. As the cleaning time is 10 min in the process currently in use, 90 min of cleaning was equivalent to 9 times the cleaning cycle. The criteria for acceptance were readings above -0.22 mm at the stencil center and above -0.22 mm on the other locations.

Stencil Quality Inspection

The two-factor three-level full factorial experimental design was used. The factors considered were taper angle and laser cutting speed. Details of the experimental design are shown in Table 1. The taper angle (Fig. 2.) was determined by Equation (1). Quality characteristics included the occurrence of clogging and aperture roundness, as in Fig. 3. Clogging was defined as the percentage of aperture area covered by the steel scrap, while roundness was determined by image measurement software. The stencil aperture design is shown in Fig. 4.

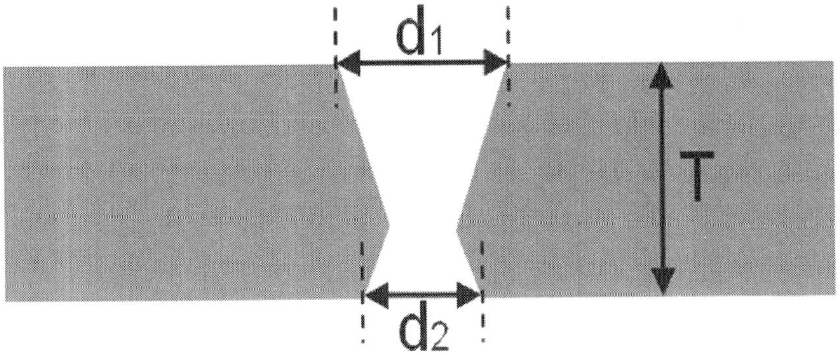

Figure 2: Taper angle.

$$\alpha = \tan^{-1}((d1-d2)/2*T) \quad (1)$$

The d1 and d2 are the aperture sizes at the entrance side and exit side, respectively; T is the stencil thickness.

Table 1: Laser cutting experimental design

Factors	Levels		
Taper Angle	2°	4°	7°
Speed(mm/s)	12	10	8

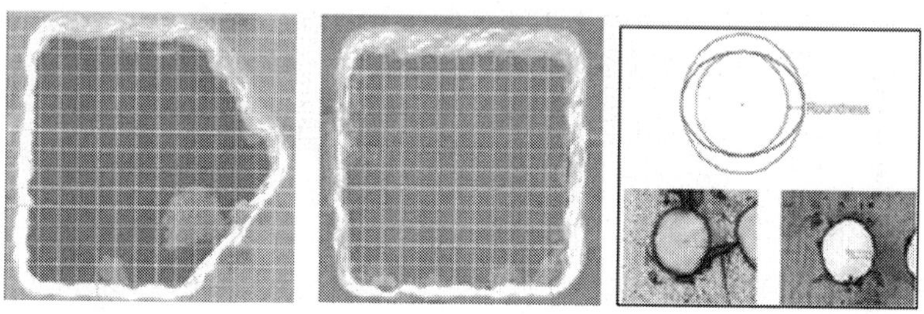

Figure 3: Clogging and aperture roundness.

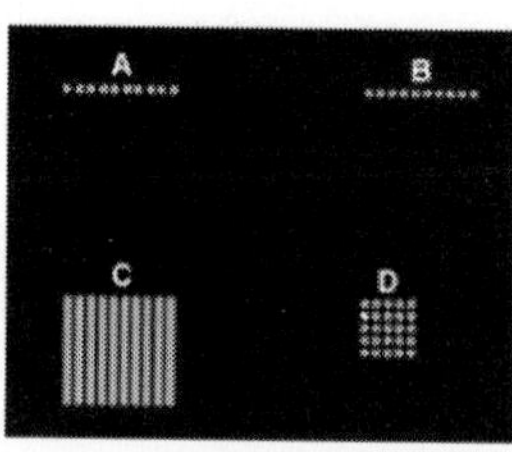

Area	Component pad	Scale (mm)	Inspection Item
A	01005 Home Plate	Side:0.186 Space:0.129mm	Clogging
B	Square	0.2mmx0.2mm	Clogging
C	Connecter	0.15mmx2.65mm	Clogging
D	BGA Pitch 0.3	Diameter:0.2mm	Roundness

Figure 4: Stencil design for stencil cutting evaluation.

Electropolishing Process

The anodic polarization chart (Fig. 5) shows the relationship between the applied voltage and the anode current density. When the voltage reached the polishing region (DE), further increases in voltage did not have a significant effect. The stencil aperture logging scenario considered three polishing times.

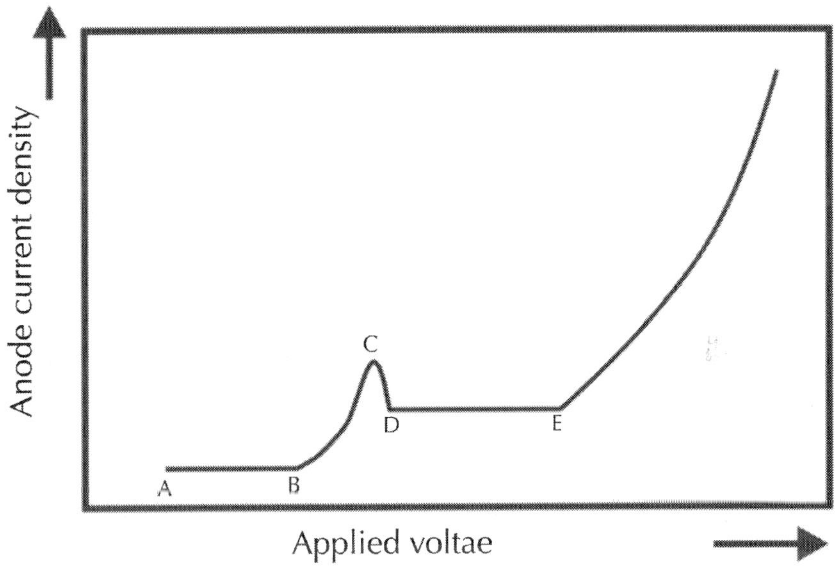

Figure 5: Anodic polarization chart.

Printing Process Assessment

Three stencil thicknesses (0.08mm, 0.1mm, 0.12mm), three post treatment methods (CP: chemical polishing, EP: electropolishing, NP: no polishing) and three taper angles (2°, 4°, 7°) were considered. Details of the experimental design are shown in Table 2. Ten stencils in total were used for the printing experiments. The squeegee printing angle remained at 60°. Stencils were cleaned after every printing. The separation velocity was 0.5 mm/sec, the printing pressure was 12 Kg and the printing velocity was 40 mm/sec. The sample size was forty, i.e., there were forty identical apertures on the stencil for each treatment. The printing was duplicated 15 times. The transfer ratio of printing was used as quality performance.

Table 2: Experimental design for printing performance evaluation

Stencil thickness	0.08				0.1					0.12
Machine	Machine A								Machine B	
Taper Angle	2°	4°	7°	4°	2°	4°	7°	4°	4°	4°
Polishing	EP	EP	EP	CP	EP	EP	EP	CP	NP	NP
Stencil ID	A	B	C	D	E	F	G	H	I	J

The test board was a bare board without a solder mask in order to minimize the potential printing variation. It is not reasonable to compare the transfer ratio when evaluating the printing performance of stencils with different thicknesses. While the amount of solder paste printed is a major factor influencing the solder joint reliability, a thick stencil with a smaller transfer ratio may still have a greater amount of solder paste deposition. Therefore, a 0.1 mm stencil was used as the standard thickness (Fig. 6). The printing performance was normalized and evaluated through the steps below.

The equation used to calculate the transfer ratio is shown below:

$$\text{volume transfer ratio} = \frac{\text{solder paste volume}}{\text{stencil thickness} * \text{aperture area}} \tag{2}$$

After ten printings, the average and standard deviation are calculated.

The stencil with the highest average printing volume is assigned a score of 10, and the second highest is assigned a score of 9....

The performance of printing variation is considered for an extra score. The top three stencils with the least printing variation (standard deviation) are assigned a score of 3, 2, 1, respectively. The results of printing for a 0.3 mm BGA are shown below.

Table 3: The results of printing for a 0.3 mm BGA

Stencil	F	I	D	A	C	J	H	E	G	B
Mean	10	9	8	7	6	5	4	3	2	1
Standard deviation		3				1			2	
Total	10	12	8	7	6	6	4	3	4	1

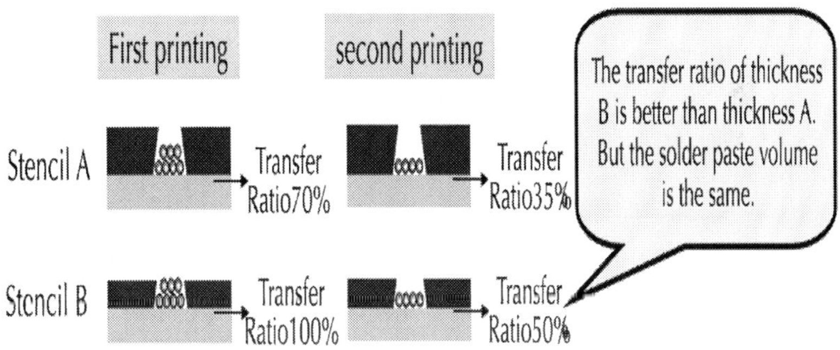

Figure 6: Plot showing difference in transfer ratio for various stencil thicknesses even with similar solder volume deposition.

Results

Material Properties Assessment

After a total of 30 hours IPA cleaning (180 cleaning cycles), separation occurred to one of screen model. It showed that the test method can be an effective evaluation of the durability of raw materials.

Stencil Quality Inspection

Results show that a taper angle of 2° and a laser speed of 8 mm/s resulted in minimal stencil aperture clogging (Fig. 7(a)). A slower laser speed was preferred for better aperture roundness (Fig. 7(b)).

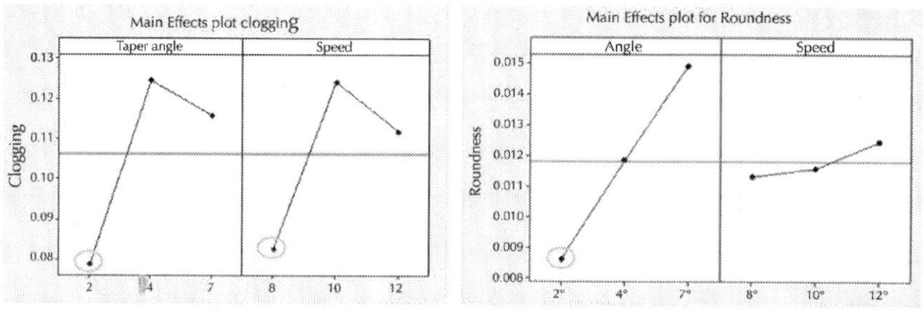

(a) (b)

Figure 7: Main effects of taper angle and laser speed on (a) clogging and (b) aperture roundness.

Electropolishing Process

Results showed that clogging was minimized by increasing the polishing time (Fig. 8). The residue was found to be carbon and oxide through scanning electron microscope/energy dispersive X-ray spectroscopy (SEM/EDS) analysis (Fig. 9). Therefore, the electropolishing process did not seem to be effective in removing the non-metal contamination and improving solder deposition. Also, chloride was detected and this may have introduced additional concerns.

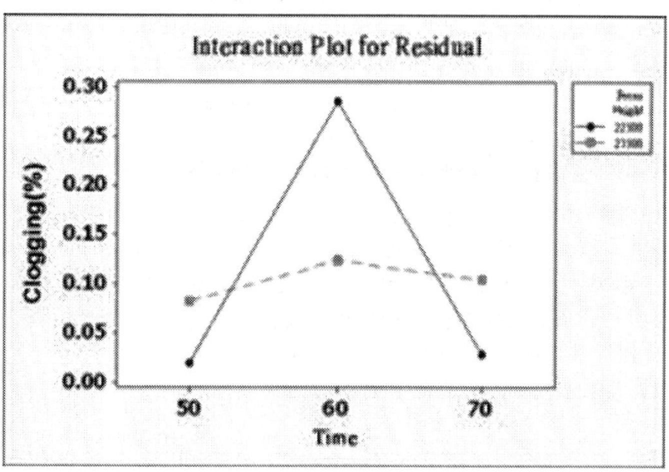

Figure 8: Main effect of polishing time on aperture clogging.

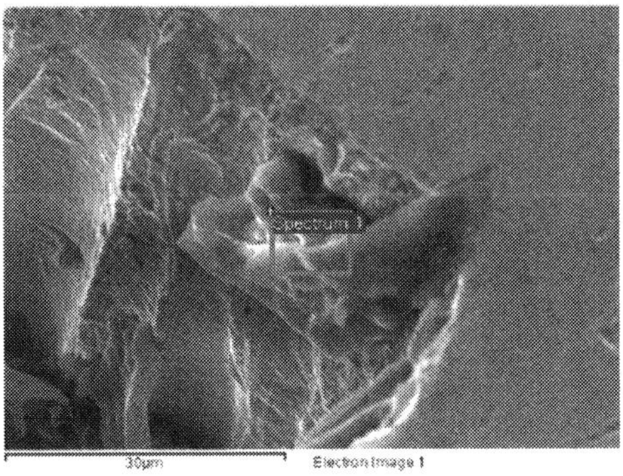

Element	Weight%	Atomic%
C K	52.48	62.64
O K	37.08	33.22
Na K	2.79	1.74
Cl K	2.94	1.19
Fe K	4.72	1.21
Totals	100.00	

Figure 9: Residue compound SEM/EDS analysis.

Printing Process Assessment

The effect of the stencil aperture shape was evaluated. For both 0.08 mm and 0.1 mm stencils, the circular aperture design was superior to the home plate shape aperture design for 01005 solder paste printing with respect to the transfer ratio (Figs. 10 and 11). For the 0.08 mm stencil, a chemical polished stencil was superior to an electropolished stencil. However, for the 0.1 mm stencil, the electropolished stencil and the stencil with no polishing performed better than the chemical polished stencil. This result was consistent with SEM/EDS analysis in that post treatment did not remove steel scraps and improve solder deposition.

The solder paste transfer ratio scores for various components are shown in Table 4. The transfer ratio for the 0201, 0.5mm BGA and 0.4mm connector were all above 100%. The stencil performance was evaluated for miniaturization applications. The components with a transfer ratio above 100% were excluded. The top three scenarios with the highest scores were (1) the electropolished stencil with a 0.1 mm thickness and a 40° taper angle; (2) the stencil with no polishing, 0.1 mm thick and a 2° taper angle; and (3) the stencil with no polishing stencil, 0.12 mm thick and a 2° taper angle.

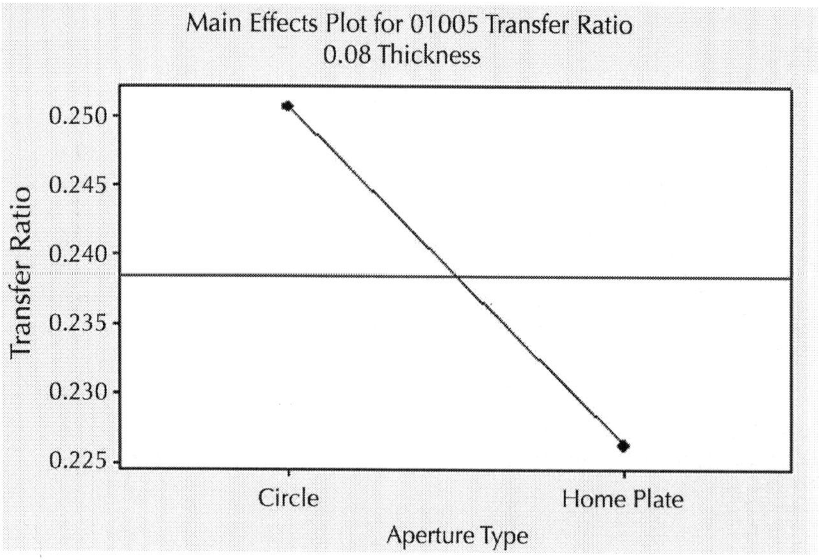

Figure 10: Influence of aperture shape on transfer ratio for stencil with thickness 0.08 mm.

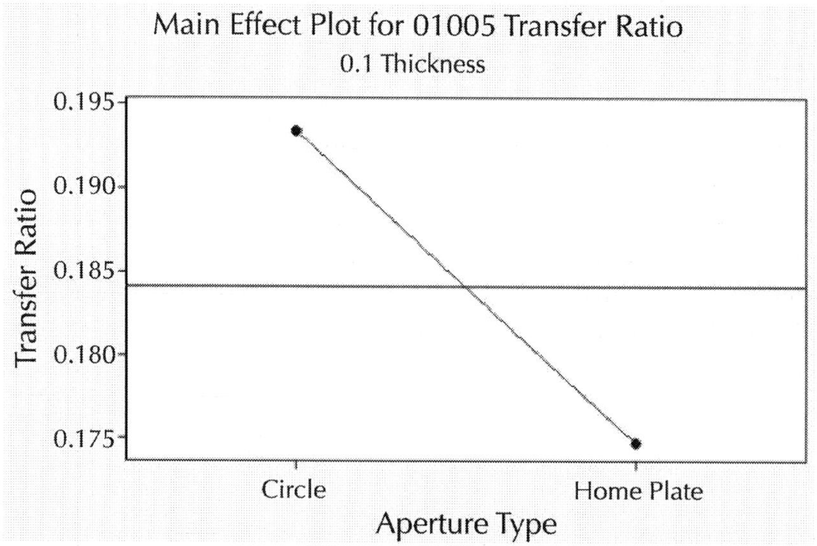

Figure 11: Influence of aperture shape on transfer ratio with thickness 0.1 mm.

Also, the 0.1 mm stencil performed better than the 0.08 mm stencil with respect to the transfer ratio. The stencil with no polishing performed better than the electropolished or chemical polished stencils.

The correlations between the transfer ratio and area ratio were investigated for the stencils with three different post treatment technologies (Figs. 12 and 13). No correlation was found for either the electropolished or chemical polished stencils. The process variation of laser machine A was more severe than that of machine B. While the industry standard IPC-7525 (IPC-7525 3.2.1.1, 2000) suggests an area ratio above 0.66, stencils with an area ratio of 0.5 still achieved a 60% transfer ratio for machine B. Also, as regards stencil manufacturing, stencils with no polishing performed better than stencils that were either electropolished or chemical polished. Significant variations in the stencil manufacturing machine were observed, which may have affected the effectiveness of the stencil evaluation. The area ratio could be used to verify the process variations of stencil manufacturing in future stencil quality assessment studies. The development of printing capability contributes to the miniaturization of electronic products with greater functionality.

Table 4: Solder paste transfer ratio score for various components

Stencil ID	F	I	J	G	A	E	D	C	H	B
01005	12.5	10.5	9	4	8	3	6	5	1	2
BGA0.4	10	10	10	6	4	3	5	4	8	1
BGA0.3	10	12	6	4	7	3	8	6	4	1
CN 0.3P	9	11	10	6	4.5	7	4	5.5	3	1
CN 0.3V	10	6	8	10	5	11	2	3.5	3.5	1
Total	51.5	49.5	43	30	28.5	27	25	24	19.5	6

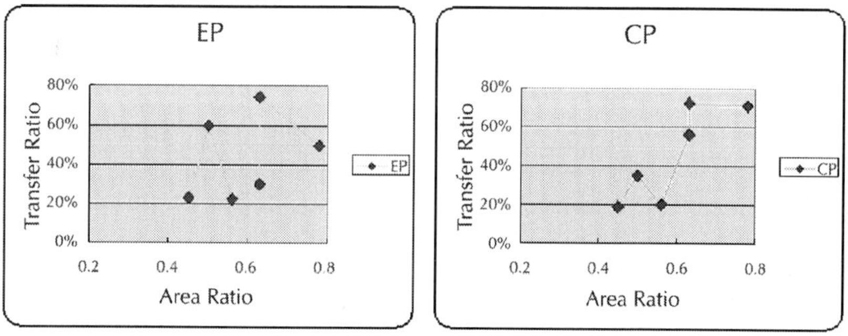

Figure 12: Correlation between the transfer ratio and area ratio for laser machine A.

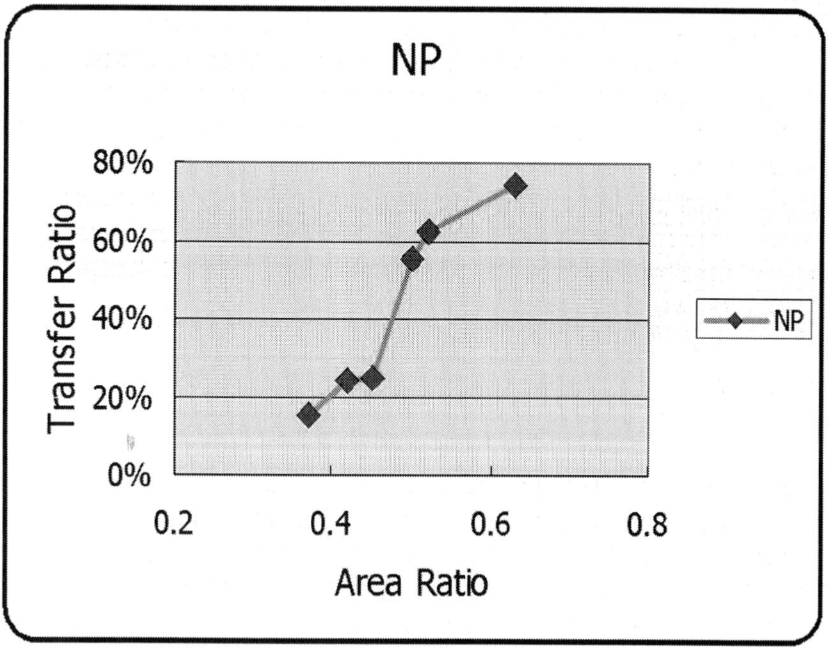

Figure 13: Correlation between the transfer ratio and area ratio for laser machine B.

CHARACTERIZATION OF REFLOW TEMPERATURE PROFILE

During reflow soldering, the solder paste melts and solidifies to form reliable solder joints. In the reflow soldering process, the PCBs are transported by the conveyor and go through the various heating zones. Prior to mass production, the process engineer has to ensure that the desired temperature is achieved by using a profile board with the same characteristics as the real product to be assembled. This is critical to ensure that the desired solder joint is formed, while at the same time taking into consideration the heat resistibility of the PCB and the electronic components. The trial and error approach is commonly used, which involves adjusting the temperature setups at various heating zones and the conveyor speed to achieve the temperature profiles on important locations of the PCB so that all comply with the process specifications [Su 1997]. The process to determine the appropriate temperature setups is time consuming as it includes the time to measure the PCB temperature, time for the thermal tracker to cool down to room temperature and the time for the reflow chamber to reach a stabilized temperature. Therefore, the schedule for production could be detained. Characterization of the reflow heating process helps to reduce setup time and the number of trials while ensuring the quality of the solder interconnections.

Temperature simulation using finite difference (FD) and computational fluid dynamics (CFD) modeling tools can achieve a high degree of accuracy, but they are complex to generate and analysis times are unacceptably long. Whalley (2004) developed a less complex approach using simplified representations of both the product and the process. Sarva and Conway (1998) abandoned mathematical modeling and utilized the concept of What-If to forecast temperature profiles. This approach, however, is appropriate primarily at the phase of product design and process planning. Su (1997) proposed a knowledge-based approach to design and implement a profile identification decision support system. Inferences through an expert system usually apply only to certain categories of products. Gao (2008) offered a convenient method to develop an oven recipe for a specific soldering profile. The method was devised to quickly achieve the proper profile shape, but the accuracy of the profiling was constrained, especially for complicated products in real applications.

Our research considered real PCBs of server applications, usually having six to eight copper layers, but sometimes up to a total of eighteen layers, which are loaded with complicated components with a large heat capacity. The algorithm suggested by Gao (2008) was used in this study as a baseline model for prediction. Data (real temperature setups) of a total of twelve server models collected from production environments were then employed to make the needed modifications to the baseline model. A 'correction value' for the baseline model was the 'response', and the temperature setting of the heating zone, conveyor speed and the total PCBA heat capacity were the independent variables. Stepwise regression analysis (SRA) was used to effectively express the relationship between the response variable and the independent variables (for each heating zone).

Preliminary Experiments

Preliminary experiments were conducted to assess the stability of the reflow oven under study and the influence of the heating zone setup on the resulting temperature profile. The effect of the PCB carrier was also considered. The three preliminary experiments considered a total of five heating zone setups (Table 5). The oven used in these experiments had nine heating zones and natural cooling (Fig. 14). The test vehicle was a bare board with eight stacking layers (Fig. 15). The length, width and thickness were 241 mm, 335 mm and 1 mm, respectively. The thickness of the copper layers on the PCB surface was 0.5 oz, and 1 oz for the inner layers. Six thermal couples were attached at four corners on the top side and center locations of both the top and bottom sides. The fixture used in this study is shown in Fig. 16.

Table 5: Heating zone setups considered in the preliminary experiments

Heating zone	Z1	Z2	Z3	Z4	Z5	Z6	Z7	Z8	Z9
The 1st preliminary experiment	180	180	180	180	180	180	180	180	180
The 2nd preliminary experiment	180	180	180	200	200	200	180	180	180

The 3rd preliminary experiment	Temp setting (1)	180	180	180	180	200	200	240	240	240
	Temp setting (2)	180	180	180	180	200	200	250	245	240
	Temp setting (3)	180	180	180	180	200	200	260	240	220

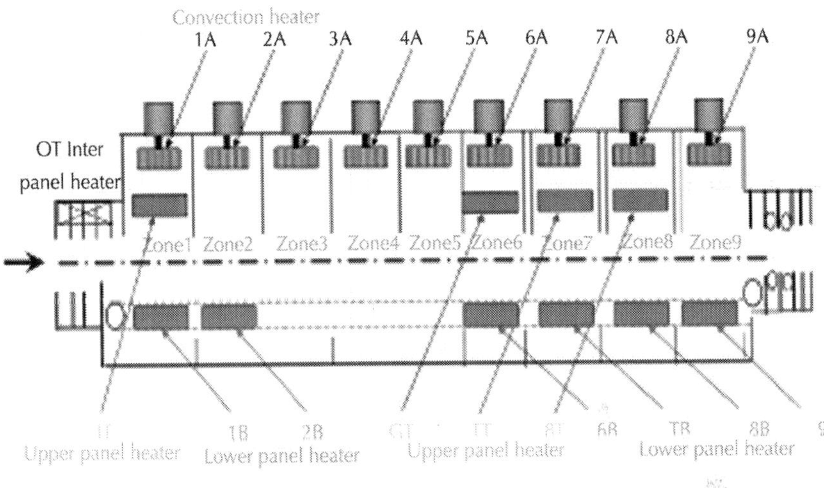

Figure 14: Reflow oven used in the preliminary experiments.

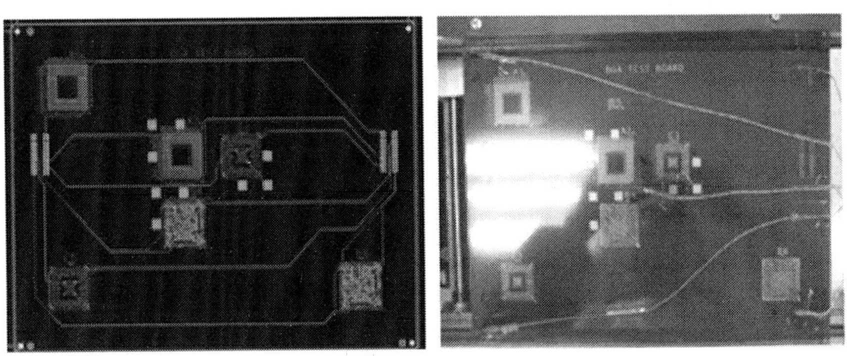

Figure 15: PCB used in the preliminary experiments.

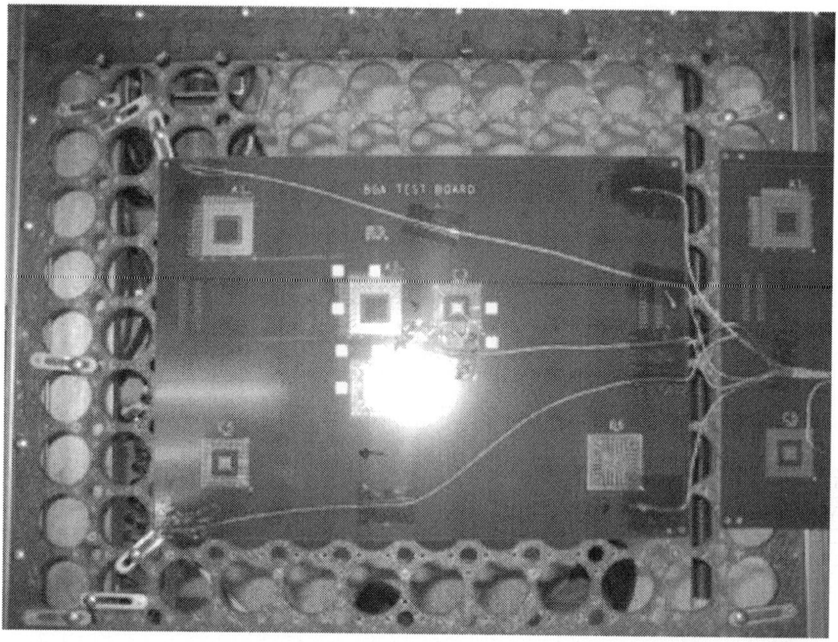

Figure 16: Fixture used in this study.

In the first experiment, the temperature setting for all heating zones was maintained at 180°C. The measured preheating and cooling temperature slopes for the PCB samples with a carrier were smaller than those for PCB samples without a carrier. The peak temperature (T_p) during the reflow zone for the PCB samples with a carrier was also lower than that for PCB samples without a carrier and with less variation (ΔT). The above mentioned observations showed the inconsistency in heating phenomena due to the carrier's heat capacity. Secondly, the measured temperatures near the adjustable side of the conveyor were higher than those near the fixed side of the conveyor. This was because the adjustable side of the conveyor was close to the center of the heater and the heating source was more stable and effective (Fig. 17). In the second experiment, the temperature setting at Z4~Z6 was 200°C and all the other zones maintained at 180°C. The results showed that the PCB temperature also increased by approximately 20°C when the temperature setting at the heating zone increased by 20°C. This indicated that the heating zone was stabilized and that the heat transfer was effective. The third experiment, in reference to Cho (2008) (Fig. 18), considered three temperature settings: (1) reflow heating zones Z7-Z9 maintained at 240°C; (2) reflow zone Z7 at 250°C and the remaining zones (Z8 and Z9) decreasing 5°C progressively; and (3) reflow zone Z7 at 260°C and the remaining zones decreasing 20°C progressively. The purpose was to investigate the effect of the temperature setting at the reflow zones on the reflow time (time above 217°C) and the peak temperature. Temperature setting (2) resulted in a higher reflow peak temperature, a longer reflow time and the least variation in peak temperature for the various PCB locations (Table 6), i.e., the reflow temperature setting decreasing 5°C progressively resulted in the desired stable (hat type) temperature profile.

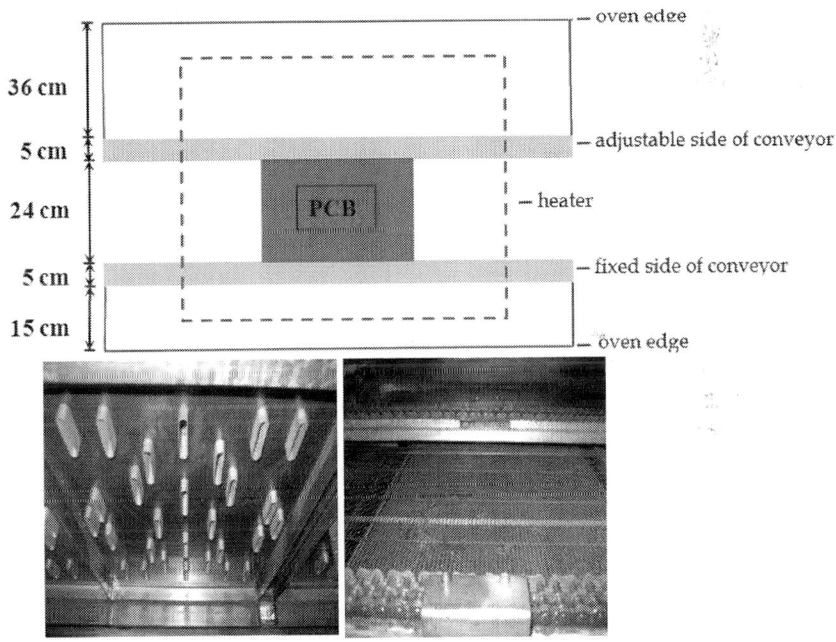

Figure 17: Plot showing location of heater and the PCB.

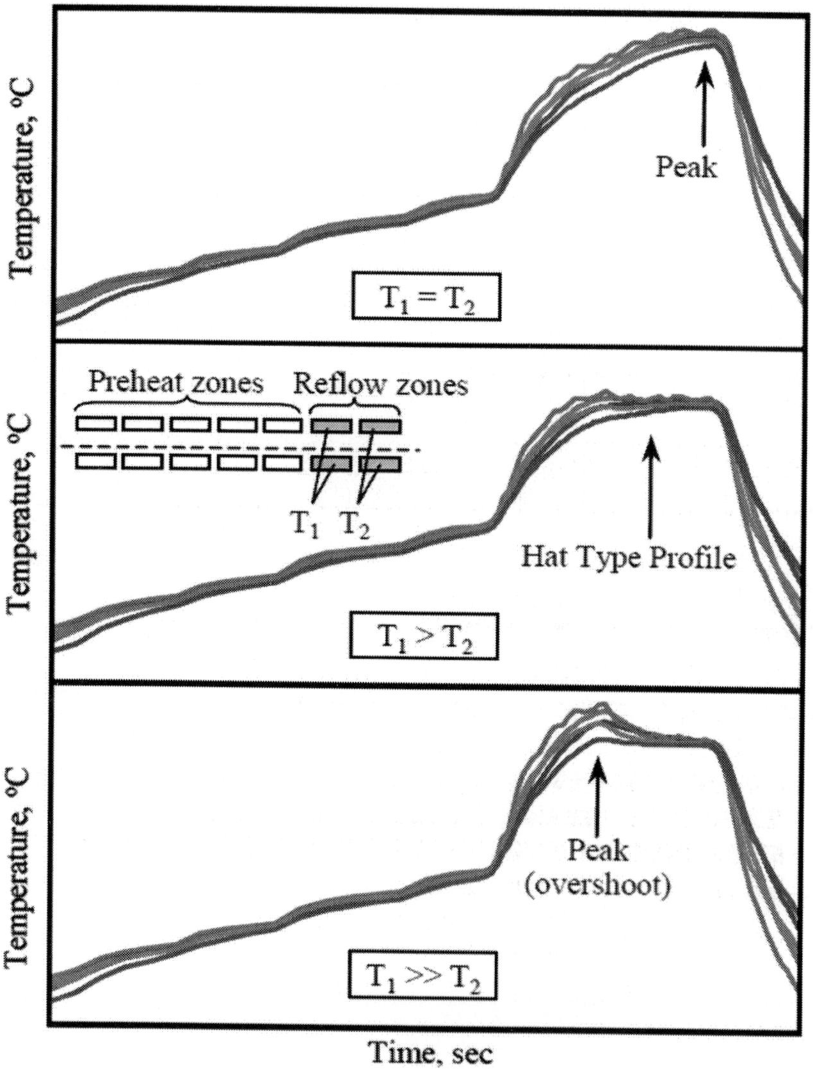

Figure 18: Temperature setting at reflow zone from Cho (2008).

Table 6: Results of the third experiment

	Temperature setting (1)		Temperature setting (2)		Temperature setting (3)	
	without carrier	with carrier	without carrier	with carrier	without carrier	with carrier
Average Tp(°C)	227	222	230	224	228	220
Reflow time (s)	60	41	71	52	68	44
ΔT (°C)	6.3	7.8	5.7	6	6.5	10.4

Reflow Process Specification

The specifications used in this study are shown in Fig. 19 and described below: (1) the temperature slope increase from 50°C to 120°C during the preheat stage should be less than 3°C/s to prevent damaging the PCB/component due to thermal shock; (2) a slow temperature rise at the soak stage reduces temperature variation across the PCB before the reflow soldering commences. A temperature rise from 150°C to 180°C should be within 60s to 120s; (3) the time above soldering material SAC305's melting point 217°C should be within 40s to 90s; and (4) rapid cooling is preferred to achieve the desired solder joint metallurgy. The temperature slope from 210°C to 170°C should not exceed 3°C/s to reduce thermal shock.

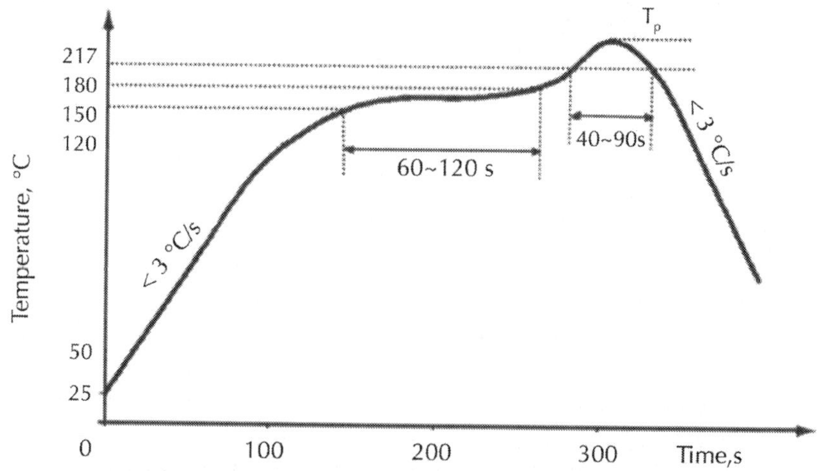

Figure 19: Reflow temperature profile process specifications.

Basic Model Suggested Temperature Settings

This study started with calculations of the suggested temperature settings based on the reflow profile prediction algorithm proposed by Gao (2008). The results were compared with the real temperature settings in the manufacturing process so that the measured PCB temperature complied with the process specifications. The result of a sample product is shown in Table 7. This task was then duplicated for a total of twelve server products.

Table 7: Basic algorithm suggested temperature settings versus real settings

heating zone	Z1	Z2	Z3	Z4	Z5	Z6	Z7	Z88	Z9	ZI0	Z11
basic algorithm	70	125	153	174	185	191	221	248	260	248	216
real settings	120	170	180	180	185	185	240	260	260	185	145
difference	50	45	22	6	0	-6	19	12	0	-63	-71

Constructing the Function of the Correction Value through Stepwise Regression Analysis

The temperature settings of twelve server products used in real manufacturing environments were collected. It is noted that the settings were determined by experienced process engineers through trial and error. This was time consuming and iterative modification efforts were required to ensure that the temperatures at critical locations of the PCB were within process specifications. This was essential before the mass-production commenced. Table 7 and Fig. 20 show the basic algorithm suggested temperature settings compared to the real settings. Real temperature settings exhibited longer soak times (in a temperature range of 150°C~180°C) and reflow times (above 217°C). Apparently, PCBs with a higher layer count and loaded with complicated electronic components were more suitable. In contrast, the basic model was only appropriate for the experimental bare board and/or low-end products. Therefore, the aim of this study was to determine the 'correction value' used to modify the suggested settings proposed by the basic algorithm and feasible for high-end products, such as a high-layer count server.

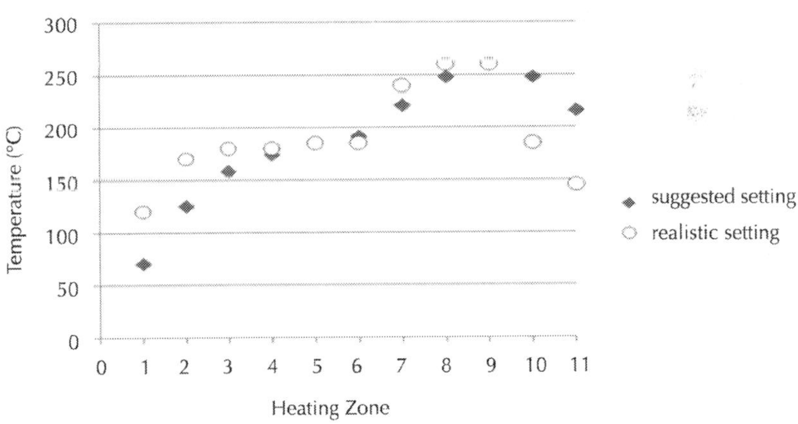

Figure 20: Basic algorithm suggested temperature settings compared to the real settings.

The correction value is considered as a function, f(x), of the product and process-related factors, namely, the total PCB assembly's (PCBA's) heat capacity, $H_{total}(m)$; the temperature setting at zone i, x_1; the temperature setting at the previous zone (zone i-1), x_2; the temperature setting at the following zone (zone i+1), x_3; and the conveyor speed, x_4. The PCBA consisted of FR4 (flame retardant-4) material, copper layer and components. The total PCBA heat capacity (J/k) is the product of the materials' mass (g) and corresponding specific heat capacities (J/g-k) (Equation 3). In this study, the mass of PCBA was measured with a digital balance. While CAD Gerber file provided information on the total volume of copper, the mass of copper was determined by multiplying the copper density (1.9 g/cm3). The components mass could thus be determined by deducting the mass of PCBA by the masses of copper and components.

$$H_{total}(m) = m_1 c_1 + m_2 c_2 + m_3 c_3 \quad (3)$$

where m1, m2 and m3 are the masses of FR4, copper and components of a given product; c_1, c_2 and c_3 are 0.6 J/g-k, 0.38 J/g-k and 0.1 J/g-k, respectively, which are the specific heat capacities of FR4, copper and components, respectively.

The function of the correction value was constructed through the data of eleven server products. Data of the other server product was then used for model verification. SRA integrates the forward regression and the backward regression. First, the forward regression brings in the item with the most prediction ability; the backward regression then removes the item(s) that are not significant (expected explanation ability) from the model. The test criteria, partial F value and type I error, , used in this study were 4 and 0.15, respectively. The cross validation continued until all the items in the complete function were tested. Statistics software Minitab was used to analyze data of the eleven server products and construct the correction value function for each heating zone. The tenth heating zone, Z10, was used for illustration (Table 8). The complete function included first power items, second power items and two way interaction items (Equation 4). The twenty items were tested for their significance, and the results showed that only item x_1^2 was brought into the function model.

$$f(x) = \beta_0 + \beta_1 g(m) + \beta_2 X_1 + \beta_3 X_2 + \beta_4 X_3 + \beta_5 X_4 + \beta_6 H(m)^2 + \beta_7 X_1^2 + \beta_8 X_2^2 + \beta_9 X_3^2 + \beta_{10} X_4^2 + \beta_{11} H(m) X_1 + \beta_{12} H(m) X_2 + \beta_{13} H(m) X_3 + \beta_{14} H(m) X_4 + \beta_{15} X_1 X_2 + \beta_{17} X_2 X_3 + \beta_{19} X_2 X_4 + \beta_{20} X_3 X_4 \quad (4)$$

Table 8: Data to construct correction value function for Z10

product	difference (°C)	total heat capacity (J: k)	Z10 set temp (°C)	Z9 set temp (°C)	Z11 set temp (°C)	Conveyor speed (cm/ min)
1	0	286.20	260	260	180	80
2	-13	485.54	240	265	160	75
3	-21	411.48	240	275	160	75
4	16	386.06	265	270	180	82
5	-7	401.34	250	270	180	75
6	-3	242.58	250	260	180	87
7	-2	316.65	250	260	150	70
8	0	191.45	260	255	180	80
9	-63	864.21	185	260	145	68
10	-62	403.69	190	265	145	80
11	11	219.52	260	260	180	73

The correction value functions for all heating zones determined by SRA are shown in Table 9. Firstly, the setting at Z1 was at 120°C for an efficient temperature rise and to prevent thermal shock to the PCBA during the preheat stage. The setting at Z9 was in the range of 260°C~275°C for the desired wetting performance and to limit the exposure of the PCBA in the elevated temperature environment during the reflow stage. Secondly, the effectiveness of the regression functions were determined by the index of model explanation ability, R-sq(adj), and model prediction ability index, R-sq(pred). The R-sq(pred) of the prediction function for the Z11 temperature setting was only 13%, and thus considered an invalid model. For heating zones Z4-Z8, no item was brought into the function with equal to 0.15. On the other hand, prediction functions for the Z2, Z3 and Z10 temperature settings were valid. Both prediction functions for Z2 and Z3 included the interaction item of 'total heat capacity', $H_{total}(m)$, and 'temperature setting'. The reason for this was as follows. Z2 and Z3 corresponded to the preheat stage of the temperature profile. PCBA at this stage had a significant heat absorption which resulted in a rapid rise in temperature. The efficiency of the heat absorption and the rise in temperature depended on the heat capacity of the PCBA; therefore, the interaction item (between 'heat capacity' and 'temperature setting') appeared as a critical item. Also, the prediction function for Z10 included only the 'temperature setting', x_1. This was because Z9-Z11 corresponded to the reflow stage of the temperature profile. The purpose of Z10 was not to raise the PCBA temperature, but to extend the reflow time duration to ensure the complete wetting of the soldering material onto the bonding pads. The PCBA's heat absorption and rise in temperature were slow; therefore, 'total heat capacity' did not play a critical role in the prediction function. Another server product was used for the verification test. The PCB temperatures were measured with the settings suggested by the baseline model and the SRA model (Table 10). The resulting temperature profiles were then compared to the process specifications. As shown in Table 11 and Fig. 21, the temperature profile resulting from the settings suggested that the SRA model was close to the target.

Table 9: Correction value functions by SRA

Heating zone	Function model	R2(adj.)	R2(pred.)
Z1	temperature setting at 120°C	-	-
Z2	y = —37.75+0.0023H(m) x1— 0.00457 H(m)x4	96%	91%
Z3	y = —27.04 — 0.00235 H(m) x3 +0.00645H(m)x4	78%	61%
Z4~ Z8	no item brought into the function	-	-
Z9	setting in the range of 260°C~275°C	-	-
Z10	y = —155.7 + 0.00241x12	86%	77%

Z11	invalid model	40%	13%

Table 10: PCB temperatures measured with settings suggested by the baseline model and the SRA model

Heating zone	Z1	Z2	Z3	Z4	Z5	Z6	Z7	Z8	Z9	Z10	Z11
baseline model	92	135	164	182	194	205	230	252	265	252	243
SRA model	(120)	162	167	182	194	205	230	252	265	250	243
ANN model	(120)	180	183	185	187	192	238	250	265	255	245

Table 11: Temperature profiles versus process specs

Temperature profile	Temp rising slope (°C /s) 50°C ~120°C	Soak time (s) 150°C~180°C	Peak temp (°C)	Reflow time (time above 217°C) (sec)	Cooling rate 210°C~170 °C (°C /s)
spec requirement	0~3	60~420	230~25	40~90	0~3
baseline model	1.38	40	234	45	1.06
SRA model	1.53	45	235	46	1.07
ANN model	1.85	72	240	62	1.51

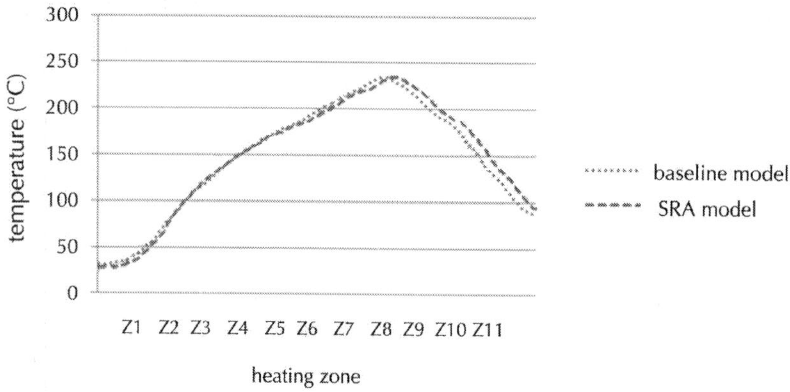

Figure 21: Temperature profiles versus process specs.

Summary

This research characterized the IR-based reflow oven and calculated the desired temperature setting through a baseline algorithm proposed in reference to twelve server samples. Correction values were as suggested by the SRA model. The results of the SRA analysis indicated that the regression functions of heating zones Z2, Z3 and Z10 were valid models with significant prediction ability. Functions of Z2 and Z3 included the item of interaction between 'total heat capacity' and 'temperature setting', while functions of Z10 included only 'temperature setting'. The results of the confirmation tests indicated that the temperature profile resulting from the settings suggested that the SRA model was close to the target.

SOLDERABILITY ANALYSIS

Solderability analysis is one of the most important characterizations for PCB assembly (PCBA), quality control and reliability (Huang et al., 2009 and 2011). The analytical techniques used for soldering properties, including visual inspection, side-view microscopy, x-ray inspection and dye staining analysis, were conducted (Huang et al., 2011 and Castello et al., 2006). The crack size percentage was classified according to the crack area. In this case, the solderability analysis was carried out to determine whether any cracks in the solder joints occurred in the CPU and RAM BGAs. The one CPU and eight RAM chips were examined to clarify the suspect ones. It may cause the failure symptom of no display to appear on an ultrabook while the power is on.

Side-View Inspection

Side-view is mainly executed to observe the surface mount components, especially the exterior row BGA solder joints in detail, non-destructively, such as foreign matter, solder joint cracks, BGA shift, missing balls and soldering related defects.

However, for an interior analysis or inspection of BGAs or PCBA components, side-view microscopy has blind spots and should be combined with other analysis methods, such as x-ray, dye staining and cross section, for further confirmation and judgment. In this study, from the side view inspection results, no abnormalities were found for the CPU1 and RAM1~8 BGAs. An illustration of the CPU1 inspection results is shown in Fig. 22. The corresponding and complementary analyses were characterized and are presented in the next sections.

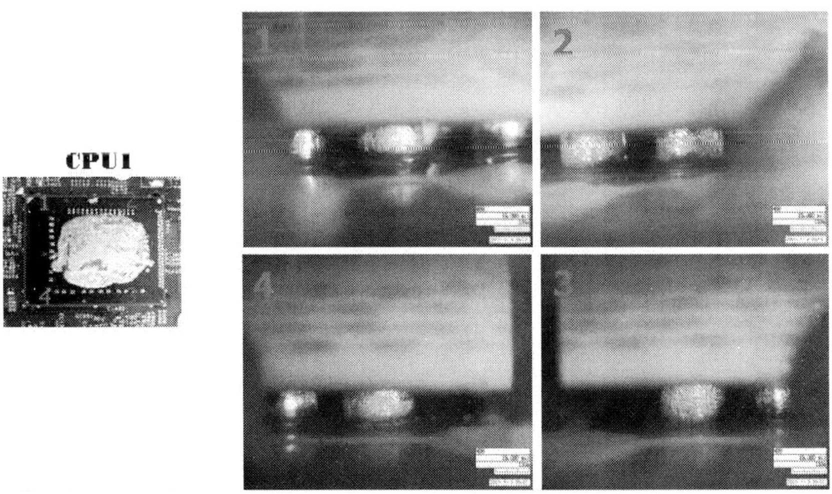

Figure 22: Side-view inspection results from the CPU1 BGA.

X-Ray Inspection

X-ray is mainly executed to examine soldering related defects for BGA, CSP and QFN, etc. X-ray inspection is regarded as a non-destructive method and is used when a high proportion of BGA solder joints is hidden under the IC package. As the internal material of BGA package shows different levels of absorbency, x-rays can determine the solder joint related defects, for example, solder bridges, missing balls, solder voids, insufficient solder or excessive solder volume. The presence and location of solder joint defects under these types of IC package may be determined by conventional 2-D X-ray in a production environment. X-ray inspection has become a common tool for soldering characterization analysis performed to verify product quality (Harrison et al., 2001).

In this study, from the x-ray inspection results, no soldering defects (short, missing balls, etc.) were observed for the CPU1 and RAM1~8 BGAs. An illustration of the CPU1 inspection results is shown inFig. 23. Voids may exist in the solder joints and make surface mount area arrays risky. According to IPC-A-610E (IPC-A-610E 8.3.12.4, 2010), the acceptance criterion for voids in solder joints is 25% or less voiding of the ball x-ray image area. In this case, void percentages were all within the inspection criterion (<25% ball x-ray image area). While some critical defects in BGA solder joints were not readily detectable by the side-view microscope and/or the x-ray inspections, dye staining analysis was carried out for more accurate soldering analysis.

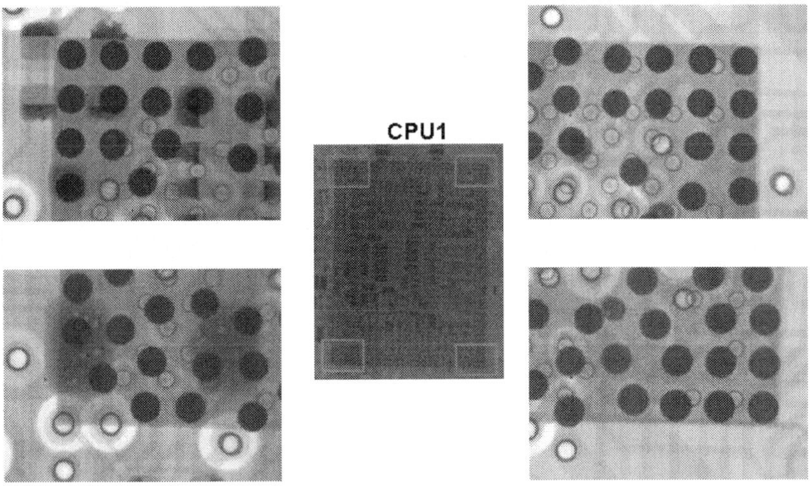

Figure 23: X-ray inspection results from the CPU1 BGA.

Dye Staining Analysis

The dye staining analysis is commonly applied to verify whether solder joint cracks have occurred under BGA packages (Liu et al., 2010 and Lau et al., 2004). The dye staining technology, which is a destructive test method used to reveal defects on the solder balls, relies on a liquid dye that penetrates existing micro cracks or under open solder balls. After the dyeing process, a heating process is used to dry the dye followed by prying off the BGA. The presence of the dye on solder balls shows the location of the failure within the solder joint and the degree of crack propagation. It is a commonplace method applied to grid array soldering problems (Huang et al., 2011).

Since RD can not identify the failure locations, dye staining analysis is performed to determine the possible causes for the no display issue when the ultrabook power is on. If the failure locations can be identified, cross section analysis is suggested to obtain more failure symptom information from other aspects.

In this study, from the dye staining analysis, solder joint cracks were found in the CPU1 (corners 2 and 3) and RAM2 (corner 4) BGAs (Figs. 24 and 25). For CPU1, the failure mode was located between PCB pad/PCB laminate (Type 5) and the crack size percentage was between

1-25%. For RAM2, the failure mode was also Type 5 and the crack size percentage was 26-50%. Cracks are one of the most important reasons for the failure symptom showing no display for the integrated chips and board in an ultrabook product.

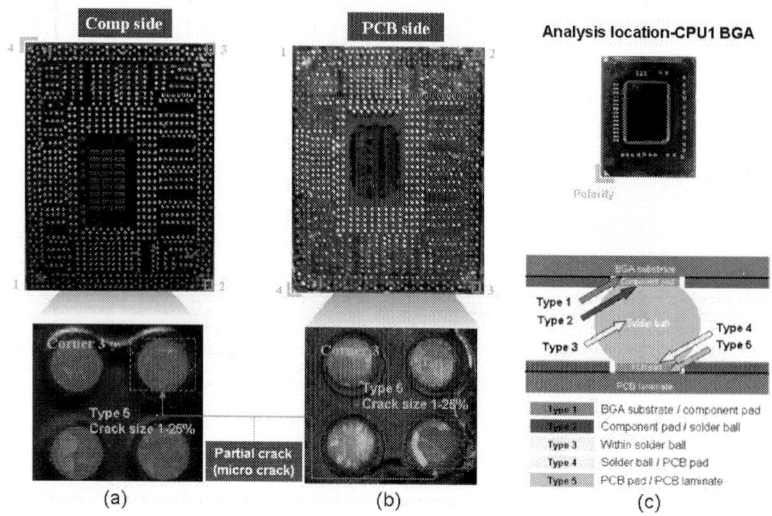

Figure 24: Dye staining analysis results from the CPU1 BGA (a) component and (b) PCB side (corner 3), (c) failure mode.

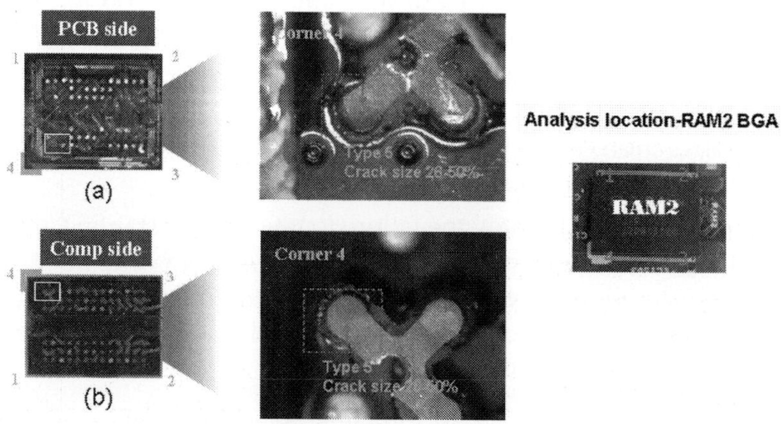

Figure 25: Dye staining analysis results from the RAM2 BGA (a) PCB and (b) component side (corner 4).

Summary

In this study, visual inspection, side-view microscopy, x-ray inspection and dye staining analysis were successfully used to characterize the soldering quality associated with microelectronics assembly. Side-view microscopy and x-ray inspection were used for a preliminary screening for failure symptoms. Consequently, dye staining analysis was used to identify the failure mode and crack size accurately. This method facilitated determining the failure causes for most soldering problems.

REFERENCES

1. Castello, T., Rooney, D. & Shangguan, D. (2006). Failure analysis techniques for lead-free solder joints. *Soldering & Surface Mount Technology*, Vol. 18, No. 4, pp. 21-27.
2. Clark, D. & Rohrer, D. (2003). Specification and qualification of machine vision technology in the SMT process, *Technical Paper, GSI Lumonics*.pp.1-8.
3. Gangue, Z. & Wang, C. (2009). Shape and fatigue life prediction of chip resistor solder joints, *International Conference on Electronic Packaging Technology & High Density Packaging (ICEPT-HDP)*.
4. Gao, J.G., Wu, Y. P., Ding H. & Wan, N.H. (2008). Thermal profiling: a reflow process based on the heating factor. *Soldering & Surface Mount Technology*, Vol. 20, No. 4, pp. 20-27.
5. Harrison, M.R., Vincent, J.H., Steen, H.A.H. (2001). Lead-free reflow soldering for electronics assembly. *Soldering & Surface Mount Technology*, Vol. 13 Iss: 3, pp. 21–38.
6. Huang, C.Y., Li, M.S., Ku, C.L., Hsieh, H.C. & Li, K.C. (2009). Chemical characterization of failures and process materials for microelectronics assembly. *Microelectronics International*, Vol. 26, No. 3, pp. 41-48.
7. Huang, C.Y., Li, M.S., Huang, S.Y., Chang, C.I. & Huang, M.H. (2011). *Material Characterization and Failure Analysis for Microelectronics Assembly Processes*. Wide Spectra of Quality Control, InTech publication, ISBN 978-953-307-683-6, Chapter 26, pp. 509-532.

8. Huang, C.Y., Lin, C.T., & Tsai, H.L. (2004). Study of Process Capability for The Stencil Printing in PCB Assembly. *Fifth Asia-Pacific Industrial Engineering and Management Systems Conference*, Gold Coast, Australia.
9. IPC-7525 3.2.1.1 (2000). Area Ratio/Aspect Ratio, pp. 3.
10. IPC-A-610E 8.3.12.4 (2010). Surface Mount Area Array – Voids, pp. 8-89.
11. Jianbiao, P., Tonkay, G.L., Storer, R.H., Sallade, R.M. & Leandri, D.J. (2004). Critical variables of solder paste stencil printing for micro-BGA and fine-pitch QFP. *IEEE Transactions on Electronics Packaging Manufacturing,* Vol. 27, No. 2.
12. Lau, J., Shangguan, D., Castello, T., Horsley, R., Smetana, J., Hoo, N., Dauksher, W., Love, D., Menis, I. & Sullivan, B. (2004). Failure analysis of lead-free solder joints for high-density packages. *Soldering & Surface Mount Technology*, Vol. 16, No. 2, pp. 69-76.
13. Liu, F., Meng, G. & Zhao, M. (2010). Experimental investigation on the failure of lead-free solder joints under drop impact. *Soldering & Surface Mount Technology,* Vol. 22, No. 3, pp. 36-41.
14. Rumelhart, D.E. & McClelland, J.L. (1986). *Parallel Distributed Processing: Explorations in the Micro-structure of Cognition.* Cambridge: MIT Press, Vol. 1.
15. Sarvar, F. & Conway, P.P. (1998). A modeling tool for the thermal optimization of the reflow soldering of printed circuit assemblies. *Finite Elements in Analysis and Design,* Vol. 30, pp. 47-63
16. Su, Y.Y., Srihari, K. & Emerson, C.R. (1997). A profile identification system for surface mount printed circuit board assembly. *Computers and Industrial Engineering,* Vol. 33, pp.377-380.
17. Tsai, T.N. (2008). Modeling and optimization of stencil printing operations A comparison study. *Computers & Industrial Engineering,* Vol. 54, pp. 374–389.
18. Whalley, D.C. (2004). A simplified reflow soldering process model. *Journal of Materials Processing Technology,* Vol. 150, pp. 134-144.
19. Yeh, I.C. (2009). Taguchi method based on neural networks and cross validation methodology. *Journal of Quality,* Vol. 14, No. 4.

Chapter 8

Computational Methods for Creation Materials with Required Composition and Structure

Konstantin Borodianskiy[1] and Michael Zinigrad[1]

[1]Laboratory for Metal and Ceramic Coatings and Nanotechnology, Ariel University Center of Samaria, Ariel, Israel

INTRODUCTION

One of the most important and complicated problems in modern industry is to obtaining materials with the required chemical composition, structure and mechanical and physical properties. Solving this problem involves great deal of time and expense, and the results obtained might be far from the optimal solution.

The development of computer technology and its accessibility have made it possible to solve problems for which there were previously unknown solutions or these methods were so tedious that they proved to be unsuitable for practical application.

There are some works where models were developed to predict the chemical composition [1, 2] and the structure [3-20] of the required deposit metal during high temperature processes, such as welding, joining, and build-up processes. These complicated models include the physical and chemical parameters of solid, liquid and gas phases, phase transition parameters, hydrodynamics' parameters, *etc*.

It is also necessary to be aware of the material's phase structure, which has a significant influence on its final properties. Predicting the phase-structure composition of a metal has been the subject of numerous papers [21-27] that included graphical representations of the phase-structure composition of the metal as a function of its chemical composition, as well as computational methods for determining the percentage of its phase.

Such a method is poorly suited to complex systems and processes described by systems of equations. In the case of mathematical modeling, the process is studied on a mathematical model using a computer, and not on a physical object. The input parameters of the mathematical model are fed into the computer, and the computer supplies the output parameters calculated in the process. The first stage in the mathematical modeling of physicochemical systems is generally the construction of thermodynamic models. This stage is very important both for ascertaining the fundamental possibility of the combined occurrence of particular chemical processes and for listing the most important thermodynamic characteristics.

In recent years mathematical modeling has been applied not only to the investigation of theoretical aspects of physicochemical processes, but also to the analysis of real technologies.

The areas of the prediction and optimization of the composition and properties of materials obtained in different technological processes are especially promising. Some of the results were obtained from the modeling of the process of the formation of a weld pool, from modeling of weld metal transformations. Important results were obtained from the studies of the physical and chemical parameters of high temperature processes, such as welding or casting, and development of kinetic model of alloy transfer. By determining the chemical composition of the weld metal the kinetic model has been developed [3-9]. Based on this model, we described the transfer of alloying elements between the slag, which is the residue left on a weld from the flux consists mostly of mixed metal oxides, sulfides and nitrides, and the metal during arc welding. The model takes into consideration the practical weld process parameters such as voltage, current, travel speed, and weld preparation geometry, and it was experimentally tested.

STRUCTURAL COMPOSITION

Structural approach is only a method development tool, *i.e.*, a means to structurize a problem, to establish connections and the order of priorities, to structurize data, *etc.*, using structural analysis.

A brief review of the major stages of structural analysis of the welding materials design problems is presented below:

- Determining the composition (structure) of the design object subject domain.
- Establishing functional relationship between the design object and the subject domain elements (direct and reverse connections).
- Establishing connections between the subject domain and the design tool (expert system).
- Determining the operational algorithm structure and the subject domain representation method for the design tool.
- Design stages structurization. Setting priorities.
- Establishing functional relationship between the design stages, tools and design object.
- Determining input and output parameters of the design tool.

This is schematically shown in Fig 1.

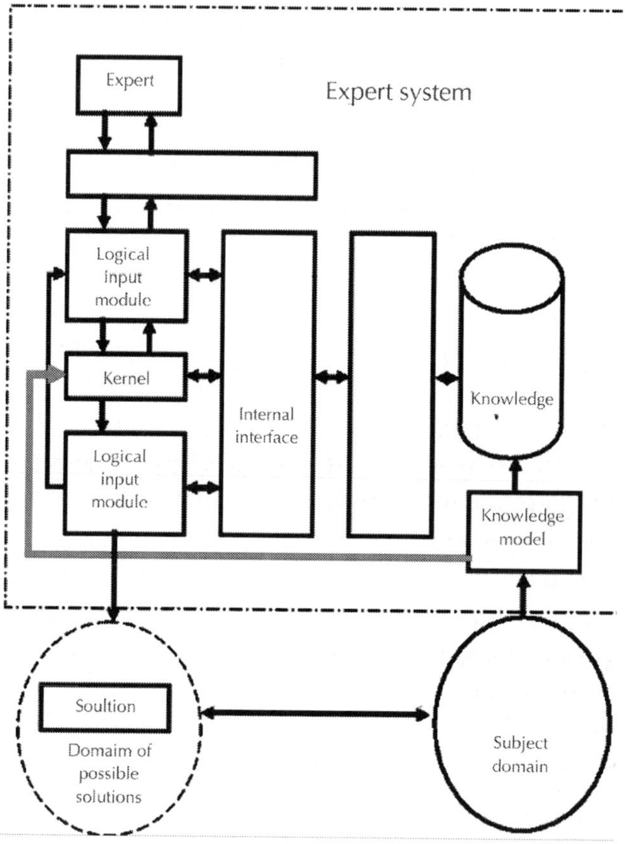

Figure 1: Simplified flow-chart of the expert system.

MATHEMATICAL MODELING OF PHASE INTERACTION IN REAL TECHNOLOGICAL PROCESS

A detailed analysis of a welded joint and its interaction with the environment are presented in the followed part. It was shown that an effective method for developing a new welding material involves solving the inverse problem of finding the formula of the material as a function of the service characteristics of the weld metal. The most important problems for the new methodology in the area of determining the electrode formula of a new welding material include devising a model of the required structure of the weld metal under service conditions and calculating the primary structure and chemical composition of the weld deposit. The chemical composition of the weld metal is determined by the initial chemical composition of the welding material and the base metal and by the nature of the physiochemical processes accompanying the interaction between the molten metal and slag.

Prediction of the chemical composition of a weld deposit and, consequently, determine its mechanical properties, is based on a kinetic analysis of the simultaneous diffusion-controlled reactions that occur between the molten metal and slag [5]. The mutual influence of the reactions and the diffusion of all the reactants in the metal and slag are also taken into account.

The analysis of the kinetics and mechanism of individually occurring reactions does not present any special difficulties at the present time and that, as a rule, its results faithfully describe the real processes. A kinetic analysis of the interaction of multi-component metallic and slag melts with consideration of the mutual influence of reactions taking place in parallel is considerably more complicated. The theoretical basis of the method consists of two assumptions:

- Under diffusion-controlled conditions the concentration ratio at the phase boundary for each reaction is close to the equilibrium value;
- The rate of transfer of the reactants to the phase boundary or away from it is proportional to the difference between their concentrations in the bulk and on the boundary of the metallic and oxide melts.

The oxidation of elements in a metallic melt can be represented by the reaction.

$$\frac{n}{m}[E_i]+(\mathrm{FeO})\rightarrow\frac{1}{m}(E_{i_n}\mathrm{O}_m)+\mathrm{Fe} \qquad (1)$$

Where E_i denotes the elements dissolved in the molten metal (Mn, Si, W, Mo, V, *etc.*), and $E_{in}O_m$ denotes the oxides in the molten slag.

A calculation of the rates of reactions of type (1) for each element does not present any difficulties. However, a separate analysis of each reaction does not correspond to the real industrial processes occurring in the weld pool. The mutual influence of both the components of the interacting molten phases and the heterogeneous reactions taking place in these complex systems must be considered. Within the approach developed, the rate v_{Ei} of mass transfer of any element (mol/cm²s) for reactions of type (1) for all the metal components with consideration of their mutual influence are defined by the expression:

$$V_{Ei} = \frac{x^m - K_{Ei}^m \times \frac{(E_{in}O_m)}{[Ei]}}{\frac{x^m}{v_{Ei}^{\lim}} + \frac{K_{Ei}^m \times (E_{in}O_m)}{[Ei] \times v_{E_{in}O_m}^{\lim}}}, \quad (2)$$

At n=1.

$$V_{Ei} = v_{Ei}^{\lim}\left[1 + b_i \times \frac{v_{Ei}^{\lim}}{4v_{Ei_2O_m}^{\lim}} - \sqrt{\left(1 + b_i \times \frac{v_{Ei}^{\lim}}{4v_{Ei_2O_m}^{\lim}}\right)^2 - 1 + b_i}\right], \quad (3)$$

Where b is defined by the equation:

$$b_i = \frac{K_{Ei}^m (Ei_2O_m)}{x^m [E_i]^2}, \quad (4)$$

At n=2.

Here K_{Ei} is the equilibrium constant of reaction (1) for the *i-th* component of the molten metal, and n and m are stoichiometric coefficients, x is the ratio between the concentration of iron oxide in the slag and the concentration of iron in the molten metal on the boundary between the interacting phases:

$$x = \frac{(FeO)'}{[Fe]'},$$
(5)

$v^{\lim}E_i$ and $v^{\lim}E_{in}O_m$ are the limiting diffusion fluxes of the components (j) of the molten metal or slag phases, $[E_i]$ and $(E_{in}O_m)$ are the initial concentrations (wt.%) of the elements and oxides in the molten phases, respectively calculated by:

$$v_j^{\lim} = \beta \times D_j^{1/2} \times C_j,$$
(6)

Where β is the mass-transfer coefficient (cm/s), D_j is a diffusion coefficient (cm^2/s), and C_j is a reagent's concentration at the phase boundary (mol/cm^3).

The rate V_{FeO} of mass transfer of iron monoxide (mol/cm^2s) which is the second reagent in reactions of type (1) defined by the expression:

$$V_{FeO} = \frac{\frac{(FeO)}{[Fe]} - x}{\frac{x}{v_{Fe}^{\lim}} + \frac{(FeO)}{[Fe] \times v_{FeO}^{\lim}}}.$$
(7)

It follows from the stoichiometry of the reaction (1) that:

$$V_{FeO} = \sum_{1}^{k} V_i = \Sigma \frac{m}{n} V_{Ei}.$$
(8)

Having substituted V_{FeO} and V_{Ei} in expression (8), we will have an equation with one unknown—*x*. Having found *x* from (2) and (3) we can find V_{Ei}.

The scheme of the analyzed technological process of fusion welding process is presented in Fig. 2.

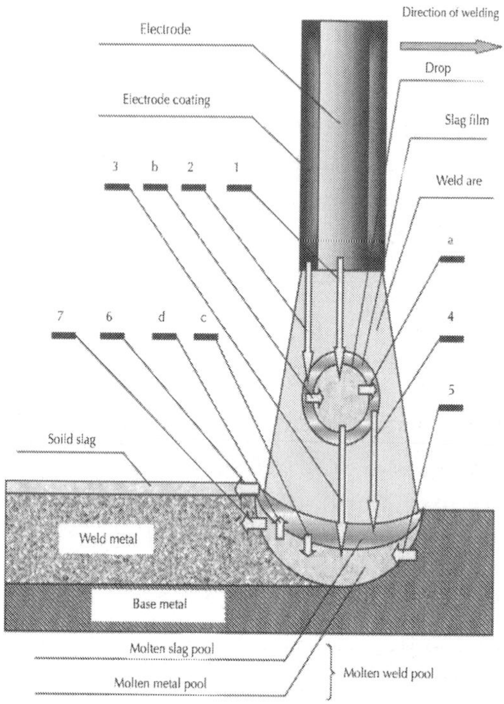

Figure 2: Scheme of the fusion welding process.

On the scheme, figures denote the direction of material transfer, and letters denote the interaction of phases:

- melting of the electrode bare and formation of a drop (1);
- melting of the electrode coating and formation of slag film over the drop (2);
- transfer of the drop metal (which has reacted with slag film at the stage of transfer) to the metal pool (3);
- transfer of the slag film (which has reacted with the drop metal at the stage of transfer) to the slag pool (4);
- melting of base metal (5);
- crystallization of the slag pool (6);
- crystallization of the metal pool (7);
- a, b—redox reaction at slag–metal boundary in a welding drop;
- c, d—redox reaction at slag–metal boundary in a welding pool.

The final composition of the drop in general case is determined by the concentrations in each of the powdered components in the flux of the flux-cored wire or in the electrode coating $[E_i]^l_{pd}$ and, accordingly, by the melting rates of these components v^l_{pd}, the concentration of each element in the metal sheath or in the electrode bare $[E_i]_b$ and its melting rate v_b, as well as by the rates of passage of the elements through the interface V_{Ei} of the drop and the slag film on its surface, which can be calculated in accordance with the methods described above. The values of v^l_{pd} and v_b are found from empirical relations as functions of the technological parameters of the process. Thus, the concentration of the *i-th* element in the drop at any moment in time *s* can be calculated from the equation:

$$[Ei]^{\tau} = \frac{v_b \cdot [Ei]_b \cdot d\tau + \sum_{l=1}^{L} v^l_{pd} \cdot [Ei]^l_{pd} \cdot d\tau + 100 \cdot M_{Ei} \cdot A_d \cdot V_{Ei} \cdot d\tau}{m^{\tau}_d}, \quad (9)$$

Where A_d is the surface area of the liquid drop, ^{l}labels the type of powder, L is the number of types, M_{Ei} molar mass of the *i-th* element, and m^{τ}_d is the mass of the metal drop at the time *s*.

The final drop composition thus calculated $[E_i]_d$ is used to calculate the concentration of the *i-th* element in the weld pool at any time *s*. The composition of the pool and therefore the composition of the weld metal are determined by the concentrations of the elements in the liquid drop $[E_i]_d$ and accordingly by the rate of descent of the drops into the liquid pool v_d, the concentration of each element in the base metal $[E_i]_{bm}$, the melting rate of the base metal v_{bm} and, as in the case of the drop stage, by the rate of passage of each element through the interface between the metal and slag pools.

In accordance with the foregoing, the expression for calculating the final composition of element *i* in the crystallized metal can be written in the form:

$$[Ei]^{\tau} = \frac{v_d \cdot [Ei]_d \cdot d\tau + v_{bm} \cdot [Ei]_{bm} \cdot d\tau + 100 \cdot M_{Ei} \cdot A_p \cdot \int_0^{\tau} V_{Ei}\, d\tau}{m^{\tau}_p} \quad (10)$$

Where A_p is the interfacial interaction area between the metal and the slag, and m^{τ}_p is the mass of the weld pool at the time *s*.

Thus, the proposed method can be used to find the chemical composition of the molten metal in the weld pool, *i.e.*, of the metal in a welded joint. This chemical composition is the starting point for determining the quantitative and qualitative composition of the phases of the weld deposit.

PHYSICOCHEMICAL ANALYSIS OF PRIMARY CRYSTALLIZATION AND CARBIDE FORMATION

The subsequent transformations of the molten metal are associated with the primary and secondary crystallization processes, *i.e.*, the phase transformations in the multi-component alloy.

After determining the chemical composition of the *i-th* component in the solid phase at the crystallization time τ, we should determine its distribution between the austenite and the carbide phases that have formed at the primary crystallization process.

Let us use the chemical composition of the liquid molten metal in the weld pool as a starting point for examining the primary crystallization process. As we know from the theory of welding processes, crystallization of the weld pool proceeds under highly non-equilibrium conditions in the absence of convective stirring of the metal in the «tail» of the weld pool, *i.e.*, at the crystallization front. Therefore, the process of distributing the components between the liquid and solid phases is controlled only by diffusion. Another important factor that determines the distribution of the components is the concentration buildup occurring at the crystallization front. These factors produce concentration-induced supercoiling, which, together, with thermal supercoiling, is responsible for the cyclic character of weld pool crystallization and the chemical non-uniformity of the crystallized weld metal. At any moment during crystallization of the weld pool, the amount of the *i-th* component that has passed from the liquid phase into the solid phase can be defined as:

$$Ei^{s} = Ei^{0}[1-(1-K_{eff})\exp(-\frac{L^{t}v_{cryst}}{D_{i}^{l}})] \tag{11}$$

where Ei^s is the concentration of the *i-th* component in the solid phase at the crystallization time т, $Ei^0$is the initial mean concentration of the *i-th* component in the molten phase, K_{eff} is the effective distribution coefficient, L^t is the distance from the crystallization starting point (the length of the crystallite at the crystallization time т), v_{cryst} is the crystallization rate, and Di^l is the diffusion coefficient of the *i-th* component in the molten phase.

After determining the concentration of the *i-th* component in the solid phase at the crystallization time*t*, we still cannot determine its distribution between the austenite and the carbide phases that have formed at the crystallization process. The factors that influence carbide formation can be divided into two groups:

Physicochemical factors, which directly determine the nature of the carbide-formation process.

Technological factors, which indirectly influence the carbide-formation process by altering the physicochemical factors parameters.

In our work, the principles governing carbide formation in an alloyed iron-carbon weld deposit were formulated on the basis of a detailed physicochemical analysis of the formation of primary carbides as compounds of carbon with *d* metals according to the quantum-chemical theories of the electronic structure of *d* metals and primary carbides. The carbide forming reaction can be described as follows:

$$xEi + yC \rightarrow (Ei)_x C_y. \tag{12}$$

According to these principles, the amount of carbon that is used to form the carbide of the *i-th* metal is proportional to the atomic radius of the metal (R_i) and is inversely proportional to the number of electrons in the *d* sublevel of the metal (d_i). We introduce the concept of the absolute Carbide Forming Ability (CFA) tendency of the *i-th d* metal (Θ_i) as the ratio:

$$\Theta_i = \frac{R_i}{d_i}. \tag{13}$$

It follows from an analysis of (12) that the carbide-forming tendency increases along the series consisting of: Fe, Mn, Cr, Mo, W, Nb, V, Ta,

Ti, Zr, and Hf, in good agreement with the results in [24, 25]. The distribution of the alloying elements and carbon between the liquid and the solid phases is given by (11). Diffusionless decomposition of the supersaturated solid solution to austenite and carbide phases occurs during crystallization. The amount of carbon bound by any carbide-forming element is determined by the stoichiometry of the compound (Me_xC_y) and can be found from the following expression:

$$E_{Ci}^{(c)} = Ei^{(c)} \frac{yA_C}{xA_i},$$

(14)

where x and y are stoichiometric coefficients, A_C and A_i are the atomic weights of carbon and the carbide-forming element, respectively, and $E_i^{(c)}$ is the concentration of the carbide-forming element in the carbide phase. For primary carbides, the value of x is always equal to 1, and y takes values from 0.4 to 1.0, depending on the homogeneity region of the respective carbide. It is logical to assume that only the portion of the alloying elements and carbon that cannot be dissolved in austenite at the respective temperature is used for carbide formation:

$$E_C^{(c)t_k} = E_C^{(s)t_k} - E_C^{(\lim)t_k}$$

(15)

Where $E_i^{(c)tk}$ is the concentration of carbon that is not dissolved in austenite, $E_i^{(s)tk}$ is the carbon concentration given by (13) at the crystallization time, and $E_i^{(\lim)tk}$ is the solubility limit of carbon in austenite at the respective crystallization temperature at the time t_k. The distribution of carbon between the carbide phases and the alloy will be proportional to the relative carbide-forming tendency of the respective transition element

$$\frac{\Theta_i}{\sum_{i=1}^{l} \Theta_i}$$

And its concentration in the alloy a_i it is now clear that the proportionality factor for the *i*-th carbide-forming element is:

$$\eta_i = \frac{1}{2}\left(\frac{a_i}{100} + \frac{\Theta_i}{\sum_{i=1}^{n}\Theta_i}\right).$$

(16)

Then the concentration of the *i-th* carbide-forming element bound in the corresponding carbide phase at the time t_{kt} can be defined as (wt.%):

$$Ei^{(c)t_k} = \eta_i \cdot E_{\mathrm{C}}^{(c)t_k} \frac{xA_i}{yA_{\mathrm{C}}},$$

(17)

And the concentration of the *i-th* carbide-forming element dissolved in austenite at the time t_k determined by (wt. %):

$$Ei^{(a)t_k} = Ei^{(s)t_k} - Ei^{(c)t_k}.$$

(18)

The concentration of carbides formed at the time t_k (wt. %) is the sum of the carbon concentration and the total concentration of the carbide-forming elements that have participated in carbide formation:

$$Q_k^{t_k} = E_{\mathrm{C}}^{(c)t_k} + \sum_{i=1}^{l} Ei^{(c)t_k}.$$

(19)

Then the austenite content (wt. %) is:

$$S^{(a)} = 100\% - Q^I$$

(20)

The mean concentrations (wt. %) of carbon and the alloying elements in the austenite phase can be found, respectively, as:

$$E_{\mathrm{C}}^{(a)} = \frac{\sum_{k=1}^{z} E_{\mathrm{C}}^{t_k}}{zS^{(a)}} 100\%,$$

(21)

$$Ei^{(a)} = \frac{\sum_{k=1}^{z} Ei^{t_k}}{zS^{(a)}} 100\%. \tag{22}$$

Thus, at the end of primary crystallization, we know the mean chemical composition of the austenite phase, as well as the quantitative and qualitative composition of the carbide phases in different zones of the formed metal. Equations (11) and (13)-(22) comprise a phenomenological model of the primary non-equilibrium crystallization of the weld pool and the formation of the weld metal. At the end of primary crystallization, we have a weld deposit of complex phase and structural composition that consists of primary carbides and of austenite phases.

PHYSICOCHEMICAL ANALYSIS OF SECONDARY CRYSTALLIZATION

Secondary crystallization is accompanied by diffusion-controlled evening of the composition of the crystallized metal to the composition specified by expressions (21) and (22), and partial coagulation of the primary carbides along their grain boundaries during cooling. When the temperature for the limiting solubility of carbon and the alloying elements in austenite is reached, the isothermal decomposition of austenite occurs, and the distribution of carbon between the carbide phases is proportional to the CFA of the respective transition element.

The evolution of the system in this stage could be predicted theoretically on the basis of the corresponding phase diagrams. However, the construction of such phase diagrams is extremely difficult for the multi-component alloys under consideration, in which the concentrations of the alloying elements can vary from several percent to tens of percent by weight. Taking into account the features of the crystallization of a weld metal noted above, we should be able to predict its phase constitution on the basis of pseudo-binary phase diagrams on the level of a qualitative estimate. There will still be a probability of a high degree of deviation from reality. This is because equilibrium phase diagrams do not take into account the real nature of the crystallization of a weld metal and the effects of the thermal-straining cycle during high temperature processes, such as welding, as well as the cyclic nature of primary crystallization, which results in chemical non-uniformity of the crystallizing metal. More than 70 years, a similar problem has been solved for certain types of molten metals in a weld pool using the phenomenological Schaeffler constitution diagram [21]. The Schaeffler diagram is a real empirical diagram that is constructed for the weld metal in the initial state after welding for ordinary averaged manual arc welding regimes are shown in Fig. 3.

Schaeffler diagram indicates a real metal microstructure formed after secondary crystallization for high alloy steel welds but can be applied for a high variety of high temperature processes in steel making processes. The microstructure composition consists of the ferrite; austenite and martensite phases can be calculated by the following equations which are called equivalents. Chromium equivalent is calculated using the weight percentage of ferrite stabilizing elements:

$$Cr_{eq} = \%Cr + \%Mo + 1.5 \cdot \%Si + 0.5 \cdot \%Nb + 0.8 \cdot \%V + 4 \cdot \%Ti, \tag{23}$$

And nickel equivalent is calculated using the weight percentage of austenite stabilizing elements:

$$Ni_{eq} = \%Ni + 30 \cdot \%C + 0.5 \cdot \%Mn + 1.6 \cdot Al + 19 \cdot N + 0.3 \cdot Cu. \tag{24}$$

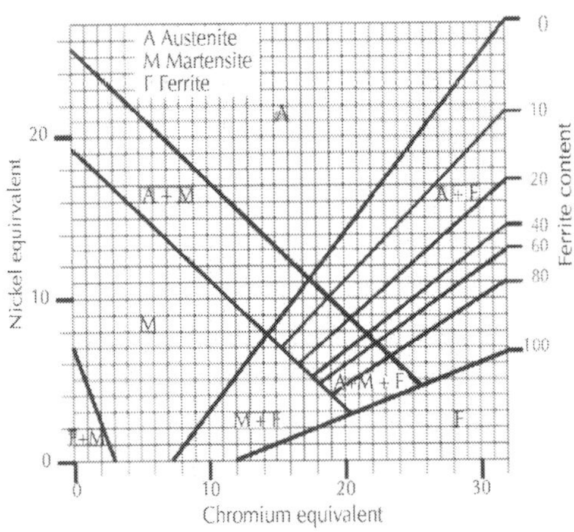

Figure 3: Schaeffler diagram.

However, when a metal contains a considerable amount of carbon (more than 0.12 wt. %) and ferrite-forming elements, are also carbides (NbC, TiC, VC, *etc.*) form in the metal microstructure. The possibility of their formation must be taken into account because the equivalent values of chromium and nickel can deviate significantly in this case from the values calculated using equations (22) and (23) proposed by Schaeffler without consideration of the formation of carbides phases.

The decomposition of austenite begins at 1100-1200 K and is accompanied by the precipitation of secondary carbides, which form mainly with chromium and iron (carbides with the general formulas Me_3C, $Me_{23}C_6$, Me_7C_3, and Me_6C). Empirical relations for predicting the type of carbide formed were determined using literature data [28, 29] and the results of our own research 4-7, 9] on the basis of the ratio between the atomic concentrations of the carbide-forming element and carbon in austenite and the parameters of the thermal-straining cycle during welding. In analogy to (17), we can write:

$$Ei^{(c)} = \sum_{j=1}^{k} w_j \, \eta_i \cdot E_C^{(d)} \frac{xA_i}{yA_C}. \tag{25}$$

Here w_j is the fraction of carbon in the *j-th* carbide phase relative to the total amount of carbon used to form carbides of the *i-th* alloying element, $E_C^{(d)}$ is the concentration of carbon in the austenite decomposition products, and η_i is the coefficient defined by (16). Then the concentrations (wt. %) of the carbide-forming element dissolved in the matrix is given by the expression:

$$Ei^{(b)} = Ei^{(a)} - Ei^{(c)}, \tag{26}$$

And of the carbon dissolved in the matrix are given by the expression:

$$E_C^{(b)} = E_C^{(a)} - \sum_{i=1}^{n}\sum_{j=1}^{k} w_j \, \eta_i \cdot Ei^{(c)} \frac{yA_C}{xA_i} \tag{27}$$

The concentration (wt. %) of the carbide phases formed as a result of secondary crystallization is:

$$Q^{II} = E_C^{(a)} - E_C^{(b)} + \sum_{i=1}^{n} Ei^{(c)}, \tag{28}$$

And the total concentration (wt. %) of the carbide phases in the weld metal is:

$$Q_{Hd} = Q^I + Q^{II}. \tag{29}$$

The concentration (wt. %) of the matrix in the weld deposit is determined from the expression:

$$S^{(b)} = 100\% - Q_{Hd}. \tag{30}$$

The concentrations (wt. %) of carbon and in the matrix is:

$$E_C^b = \frac{E_C^{(b)}}{S^{(b)}} \cdot 100\%, \tag{31}$$

And of the alloying elements in the matrix is:

$$Ei^b = \frac{Ei^{(b)}}{S^{(b)}} 100\% \tag{32}$$

Equations (24)-(31) comprise a phenomenological model of the secondary crystallization process in the weld metal which enables us to predict the phase constitution of the weld deposit.

TECHNOLOGICAL EXPERIMENTS USING MODELING APPROACH

The task of the following step is developing of a new flux-cored wire for forming of build-up layer with shock-abrasion resistance properties. The inverse problem of flux-cored wire computation has been solved (using mentioned model) which can provide us with the required chemical composition of build-up metal and as the result it can give us the required mechanical properties. These required properties are achieved by austenite-martensite matrix structure with 10 wt. % of carbides uniformly distributed in it.

Austenite has a FCC structure allows holding a high proportion of carbon in its solution. In our case austenite is used for shock resistance thanks to its impact energy absorbance ability. Martensite has a BCT structure where the carbon atoms constitute a supersaturated solid solution and as a result it has the hardest and strongest properties. Therefore martensite serves as abrasion resistant according to its mechanical properties. The stable carbide phase brings the better toughness and additional abrasion resistance and also ensures uniform distribution of the hardness properties. Under intensive impact loading, some amount of metastable austenite absorbs part of the impact energy and transforms into addition martensite phase.

The mathematical model permits prediction of the composition of the weld metal as a function of the compositions of the starting materials and the technological parameters of the welding process. Prediction of the microstructure of the weld metal is based on computer simulation of a Schaeffler diagram and the process of carbide formation in steels.

A cold-rolled ribbon (1008 steel) was filled with a powder mixture calculated using the mentioned model. The main alloying elements, in final wire, were: graphite, ferrotitanium, chromium and nickel powders. From Hume-Rothery rules it is known that the crystal structures of the solute and the solvent must be the same. Here the mentioned alloying elements should be dissolved in FCC structure (austenite phase). It is also known that the size difference between solute and solvent must be < ~15%.

Austenite and carbide are the only phases crystallize during primary crystallization process. Chromium and nickel dissolves well in the austenite formed matrix. However, titanium, because of its high difference in atomic radii as compared to γ-iron, and because of its different lattice structure, poorly dissolves in austenite. Some amount of non-dissolved in austenite chromium and titanium forms carbides.

By the end of the secondary crystallization, the stable carbide phases will be stay and residual amount of the alloying metals will be dissolved in the metal matrix.

The required properties for shock-abrasion resistance were the input, and the output is the needed alloying elements and their wt. % of the flux and wt. % of the final wire. The output is presented in Table 1 which presents chemical composition of the base metal (A516), wire band (cold-rolled ribbon, 1008 steel) and alloying elements the band was filled with.

Table 1: Flux cored wire for shock-abrasion resistance output

Component	**Density (g/cm^3)**		**Core composition (quantity in 100 kg of FCE, kg)**		**Relation in the dray mixture of the flux (%)**			
FeCrC	3.48		11.17		41.86			
FeTiC	3.15		8.13		30.49			
Ni powder	2.96		4.47		16.74			
Ca$_{F}$2	1.39		2.91		10.92			
	Composition of the base materials and the build-up layers, wt%							
	C	Si	Mn	Cr	Mo	Ti	Fe	Ni
A 516	0.28	0.30	1.00	0.00	0.00	0.00	98.35	0.00
Electrode bend	0.08	0.03	0.50	0.12	0.10	0.01	98.82	0.25
Required weld	1.00	0.60	0.80	5.50	0.00	2.00	86.57	3.50
Layer 1	0.96	0.48	0.54	5.30	0.01	2.00	86.32	3.48
Layer 2	1.11	0.53	0.42	6.71	0.13	2.54	83.13	4.41
Layer 3	1.19	0.54	0.39	7.09	0.13	2.68	82.28	4.65

Build-up samples were produced using flux cored wire with diameter 1.7 mm by process which was performed by welding machine Kemppi FU 30, PS 3500. The samples were prepared by 3 layers build-up metal. The technological parameters of the process were:

- Current: 250A.
- Voltage: 35V.
- Feed speed: 180 m/h.
- Travel speed: 30 m/h.
- Polarity: Reverse.

The samples were tested on home-made shock-abrasion resistance measuring device. The samples were subjects of the mechanic impact loading with the simultaneous continuous sea sand strew (85 g/min) of the 60 Mesh.

A 12.5 mm thickness sheet of low carbon steel, A516, consisting of 73% ferrite and 27% pearlite phase microstructure, was used as the base metal. The obtained surface (3-rd layer build-up metal) consists from martensite and austenite mixed phase microstructure and carbide stable phase, investigated by Scanning Electron Microscopy (SEM) and shown on Fig. 4.

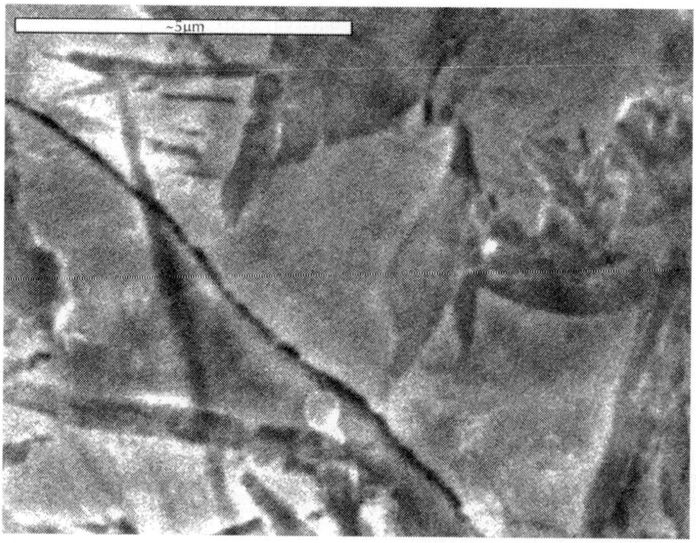

Figure 4: SEM image of surface build-up metal microstructure.

Chemical analysis of the matrix was made with Energy Dispersive X-ray Analysis (EDS) and the received results are shown in Table 2 and compared with the calculated, using the described model, results.

Table 2: Chemical composition of the weld obtained by EDS analysis and model calculation

Element	**EDS results (wt. %)**	**Calculated results (wt. %)**
C	0.97	1.19
Si	0.49	0.54
Mn	0.40	0.39
Cr	7.26	7.09
Fe	84.01	82.28
Ni	3.77	4.65
Ti	3.10	2.68

A little differences as seen in Table 2, caused as the result of technique limitation. The detected chemical elements must be rounded to 100% and this limitation doesn't take into account some chemical inclusions that usually found in steels.

As we see from the presented results, the calculated and the real chemical composition results are closed, that emphasize the correctness of the calculations performed by the model.

Hardness tests were made by Rockwell Hardness Tester. The hardness obtained on the 3-rd layer of build-up metal was 56 HRC as compare to the hardness of the base A516 metal - 90 HRB.

The difference between base metal and build up-metal for shock-abrasion resistance results obtained by shock-abrasion tests using special home-made device are presented in the plot shown in Fig. 5. The results of tested specimens presented as the plot of the weight loss per shocked area as a function of time:

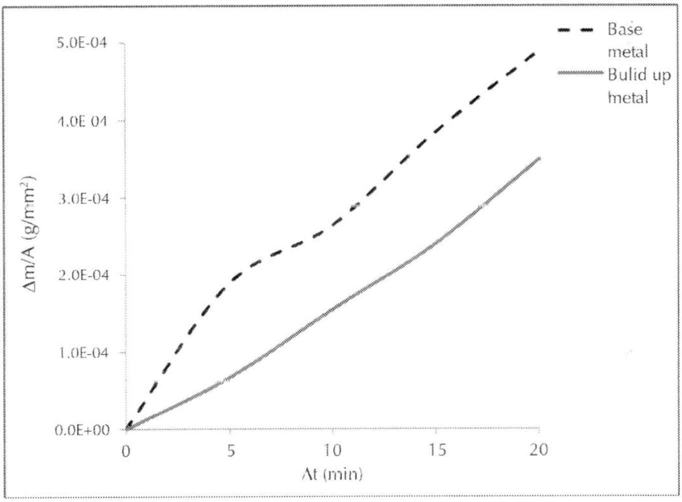

Figure 5: Shock-abrasion test results as a function of time.

It is seen from the plot that the build-up metal has an improved shock-abrasion resistance. It is found that efficiency increased by 29%, what is in a good agreement with the declared aim of the work.

CONCLUSIONS

A phenomenological model for predicting the chemical composition and structure of the non-equilibrium primary and secondary crystallization which takes into account the carbide phase formation has been developed. This allows the realization of the quantitative prediction of the metal structure and its mechanical properties.

The difficulty which was solved in the purposed model is a kinetic analysis of the interaction of multi-component metallic and slag melts with consideration of the mutual influence of reactions taking place in parallel.

Full-scale testing and investigation have been performed. The obtained results have been analyzed and treated to verify the equivalence of the model. The model was tested for solving the real technological problem, which is developing a new flux-cored wire for forming a build-up layer with shock-abrasion resistance properties. The experimental results confirm the adequacy of the calculations using this model.

REFERENCES

1. U. R Kettner, The Thermodynamic Modelling of Multi-component Phase Equilibria. Journal of the Minerals, Metals and Materials Society 199749121419
2. I Ansara, B Sundman, P Willemin, Thermodynamic Modelling of Ordered Phases in the Ni-Al System. Acta Metallurgica 1988364977982
3. M Zinigrad, V Mazurovsky, K Borodianskiy, Physico-chemical and Mathematical Modeling of Phase Interaction Taking Place during Fusion Welding Processes. Material Science and Engineering Technology 20053610489496
4. M Zinigrad, Computational Methods for Development of New Welding Materials. Computational Material Science 2006374417424
5. K Borodianskiy, V Mazurovsky, M Zinigrad, a Gedanken, Creation of Shock-Abrasion Resistance Build-up Metal Using a Physicochemical Model of High-Temperature Processes. Israel Journal of Chemistry 2007
6. K Borodianskiy, V Mazurovsky, A Gedanken, M Zinigrad, Developing a Required Structure of Metals Using Computational Methods. Material Science A 2008

7. V Mazurovsky, M Zinigrad, L Leontiev, V Lisin, K Borodianskiy, QuantitativeEstimationofCarbide-FormingAbilitiesofd-elements: proceedings of the Jubilee Scientific Conference on Physical Chemistry and Technology in Metallurgy, 2005Yekaterinburg, Russia, 8998
8. M Zinigrad, V Mazurovsky, K Borodianskiy, Physico-Chemical and Mathematical Modeling of Phase Interaction taking place during Fusion Welding Processes: proceedings of the 1-st International conference on Diffusion in Solids and Liquids, 2005Aveiro, Portugal 2841850
9. V Boronenkov, M Zinigrad, L Leontiev, E Pastukhov, M Shalimov, S Shanchurov, Phase Interaction in the Metal-Oxide Melts-Gas System. Springer, 2012
10. J Elmer, T Palmer, W Zhang, B Wood, DebRoy T. Kinetic Modeling of Phase Transformations Occuring in the HAZ of C-Mn Steel Welds Based on Direct Observation. Acta Materialia 2003511233333349
11. W Zhang, J Elmer, DebRoy T. Kinetic of Ferrite to Austenite Transformation During Welding of 1005 Steel. Scripta Materialia 20024610753757
12. M Takahashi, H. K. D. H Bhadeshia, Model for Transition from Upper to Lower Bainite. Material Science and Technology 199067592603
13. S Babu, E Specht, S David, E Karapetrova, P Zschack, M Peet, H. K. D. H Bhadeshia, In-situ Observation of Lattice Parameter Fluctuations in Austenite and Transforamtion to Bainite. Metallurgical and Materials Transaction A 2005361232813289
14. S Babu, S David, J Vitek, K Mundra, DebRoy T. Development of Macro- and Microstructures of Carbon-Manganese Low Alloy Steel Welds: Inclusion Formation. Material Science and Technology 1995112186199
15. Ø Grong, Metallurgical Modelling of Welding. The Institute of Materials Publishing, 2nd ed., London, UK, 1997
16. T Skrzypszak, E Wegrzyn-skrzypczak, Mathematical and Numerical Model of Solidification Process of Pure Metals. International Journal of Heat and Mass Transfer 2012554276

17. H Assadi, a Phase-Field Model for Non-Equilibrium Solidification of Intermetallics. Acta Materialia 2007555225
18. S Serajzadeh, a Mathematical Model for Prediction of Austenite Phase Transformation. Materials Letters 2004581597
19. G Krauss, Steels Processing, Structure, and Performance. ASM International, Ohio, USA, 2006
20. J. R Davis, Alloying Understanding the Basics. ASM International, Ohio, USA, 2003
21. A. L Schaeffler, Constitution Diagram for Stainless Steel Weld Metal. Metal Progress 19495611680680b.
22. L Beres, Proposed Modification of Schaeffler Diagram for Chrome Equivalent and Carbon for More Accurate Prediction of Martensite Content. Welding Journal 1998777273276
23. D. J Kotecki, T. A Siewert, Constitution Diagram for Stainless Steel Weld Metals. Welding Journal 1992715171178
24. T Takalo, N Suutala, T Moisio, Austenitic Solidification Mode in Austenitic Stainless Steels Welds. Metallurgical and Materials Transaction A 197910811731181
25. T Takalo, N Suutala, T Moisio, Ferritic-Austenitic Solidification Mode in Austenitic Stainless Steel Welds. Metallurgical and Materials Transaction A 1980115717725
26. S. R Chen, H. A Davies, W. M Rainforth, Austenite Phase Formation in Rapidly Solidified Fe-Cr-Mn-C Steels. Acta Materialia 1999471845554569
27. E Johnson, L Gråbæk, a Johansen, Sarholt Kristensen L. Microstructure of Rapidly Solidified Steel. Mater. Sci. Eng. 198898301
28. S. T Oyama, Chemistry of Transition Metal Carbides and Nitrides. Chapman & Hall, Glasgow, 1996
29. J Grossman, a Mizel, M Côté, M Cohen, S Louie, Transition Metals and Their Carbides and Nitrides: Trends in Electronic and Structural Properties. Physical review B. 199960963436347
30. D. V Suetin, I. R Shein, A. L Ivanovskii, Stractural, Electronic and Magnetic Properties of η Carbides (Fe3W3C, Fe6W6C, Co3W3C and Co6W6C) from first principles calculation. Physica B 20094043544

31. J Feng, B Xiao, R Zhou, Y. H Jiang, Q. H Cen, Calculation and Simulation for the Mechanical Properties of Carbides and Borides in Cast Iron. Procedia Engineering 201231676

Citations

CHAPTER 1

Jing Ma, Xiaoyong Zhang, Shixia Cheng, and Huilin Gao, "Influence of Partitioning Process on the Microstructure and Mechanical Properties of High Deformability Oil-Gas Pipeline," Advances in Mechanical Engineering, vol. 2014, Article ID 869481, 11 pages, 2014, doi:10.1155/2014/869481.

CHAPTER 2

Bo Madsen and E. Kristofer Gamstedt, "Wood versus Plant Fibers: Similarities and Differences in Composite Applications," Advances in Materials Science and Engineering, vol. 2013, Article ID 564346, 14 pages, 2013. doi:10.1155/2013/564346.

CHAPTER 3

Alicia E. Ares and Carlos E. Schvezov, "Columnar-to-Equiaxed Transition in Metal-Matrix Composites Reinforced with Silicon Carbide Particles," Journal of Metallurgy, vol. 2013, Article ID 628495, 12 pages, 2013. doi:10.1155/2013/628495.

CHAPTER 4

Nana Wang and Jinxiang Liu, "Effect of Process Parameters on Gas Nitriding of Grey Cast Iron," Advances in Materials Science and Engineering, vol. 2013, Article ID 217848, 6 pages, 2013. doi:10.1155/2013/217848.

CHAPTER 5

Jairo Tronto, Ana Cláudia Bordonal, Zeki Naal and João Barros Valim (2013). Conducting Polymers / Layered Double Hydroxides Intercalated Nanocomposites, Materials Science - Advanced Topics, Prof. Yitzhak Mastai (Ed.), ISBN: 978-953-51-1140-5, InTech, DOI: 10.5772/54803.

CHAPTER 6

V. P. Zlomanov, A.M. Khoviv and A.Ju. Zavrazhnov (2013). Physicochemical Analysis and Synthesis of Nonstoichiometric Solids, Materials Science - Advanced Topics, Prof. Yitzhak Mastai (Ed.), ISBN: 978-953-51-1140-5, InTech, DOI: 10.5772/54815.

CHAPTER 7

Chien-Yi Huang, Chen-Liang Ku, Hao Chun Hsieh, Ming-Shu Li, Chia-Hsien Lee and Cheng-I Chang (2013). Materials Assessment and Process Characterization for Lead- Free Soldering, Materials Science - Advanced Topics, Dr. Erika Hodulova (Ed.), ISBN: 978-953-51-1140-5, InTech, DOI: 10.5772/51392.

CHAPTER 8

Konstantin Borodianskiy and Michael Zinigrad (2013) Computational Methods for Creation Materials with Required Composition and Structure, Materials Science - Advanced Topics, Prof. Yitzhak Mastai (Ed.), ISBN: 978-953-51-1140-5, InTech, DOI: 10.5772/54794.

Index